T0210599

Biologically Inspired Design

A Primer

Synthesis Lectures on Engineering, Science, and Technology

Each book in the series is written by a well known expert in the field. Most titles cover subjects such as professional development, education, and study skills, as well as basic introductory undergraduate material and other topics appropriate for a broader and less technical audience. In addition, the series includes several titles written on very specific topics not covered elsewhere in the Synthesis Digital Library.

Strategic Cost Fundamentals: for Designers, Engineers, Technologists, Estimators, Project Managers, and Financial Analysts
Robert C. Creese
2018

Concise Introduction to Cement Chemistry and Manufacturing
Tadele Assefa Aragaw
2018

Data Mining and Market Intelligence: Implications for Decision Making
Mustapha Akinkunmi
2018

Empowering Professional Teaching in Engineering: Sustaining the Scholarship of Teaching
John Heywood
2018

The Human Side of Engineering
John Heywood
2017

Geometric Programming for Design Equation Development and Cost/Profit Optimization (with illustrative case study problems and solutions), Third Edition
Robert C. Creese
2016

Engineering Principles in Everyday Life for Non-Engineers
Saeed Benjamin Niku
2016

A, B, See... in 3D: A Workbook to Improve 3-D Visualization Skills
Dan G. Dimitriu
2015

The Captains of Energy: Systems Dynamics from an Energy Perspective
Vincent C. Prantil and Timothy Decker
2015

Lying by Approximation: The Truth about Finite Element Analysis
Vincent C. Prantil, Christopher Papadopoulos, and Paul D. Gessler
2013

Simplified Models for Assessing Heat and Mass Transfer in Evaporative Towers
Alessandra De Angelis, Onorio Saro, Giulio Lorenzini, Stefano D'Elia, and Marco Medici
2013

Biologically Inspired Design: A Primer

Torben A. Lenau and Akhlesh Lakhtakia

ISBN: 978-3-031-00963-1   paperback
ISBN: 978-3-031-02091-9   ebook
ISBN: 978-3-031-00163-5   hardcover

DOI 10.1007/978-3-031-02091-9

A Publication in the Springer series
SYNTHESIS LECTURES ON ENGINEERING, SCIENCE, AND TECHNOLOGY

Lecture #14
Series ISSN
Print 2690-0300   Electronic 2690-0327

Biologically Inspired Design

A Primer

Torben A. Lenau
Danmarks Tekniske Universitet

Akhlesh Lakhtakia
The Pennsylvania State University

SYNTHESIS LECTURES ON ENGINEERING, SCIENCE, AND TECHNOLOGY #14

ABSTRACT

As the existence of all life forms on our planet is currently in grave danger from the climate emergency caused by *Homo sapiens*, the words "sustainability" and "eco-responsibility" have entered the daily-use vocabularies of scientists, engineers, economists, business managers, industrialists, capitalists, and policy makers. Normal activities undertaken for the design of products and systems in industrialisms must be revamped. As the bioworld is a great resource for eco-responsible design activities, an overview of biologically inspired design is presented in this book in simple terms for anyone with even high-school education.

Beginning with an introduction to the process of design in industry, the book presents the bioworld as a design resource along with the rationale for biologically inspired design. Problem-driven and solution-driven approaches for biologically inspired design are described next. The last chapter is focused on biologically inspired design for environment.

KEYWORDS

bioinspiration, biomimicry, biomimetics, bioreplication, bionik, bionics, nature-inspired design, circular economy, contraindicated performance, design for environment, eco-efficiency, engineered biomimicry, multifunctionality, sustainability

Dedicated to sustainable societies

Contents

Preface

This primer on biologically inspired design (BID) was initiated during a sabbatical semester spent by Akhlesh Lakhtakia at Danmarks Tekniske Universitet (DTU) during the second half of 2019, at the invitation of Torben A. Lenau. The close collaboration between both of us resulted not only in the descriptions of BID approaches and the case stories required to make the reading of this book interesting to undergraduate students enrolled for BID courses, but it also made a collaboration possible with Daniela C. A. Pigosso and Tim C. McAloone for grafting BID onto *design for environment*. The combination of the two design foci makes it possible to tap into the enormous knowledge bank that the bioworld represents and apply well-proven solutions in the quest to secure sustainable societies and ecosystems on our planet.

Torben A. Lenau started in 2009 to teach BID to engineering students at DTU. More than 400 students have marched through the course since then. The course is focused on the problem-driven approach to BID illustrated by around a hundred case studies.

The solution-driven approach to BID complements the problem-driven approach. Both are treated in two chapters in this book. They are described in sufficient detail to allow practitioners as well as students to follow and apply the approaches to their own BID activities. As this book explains BID in simple terms for anyone with even high-school education, we hope that not only engineering and design students but also members of the general public interested in sustainability will profit from the time they will spend on reading this primer on BID.

Torben A. Lenau and Akhlesh Lakhtakia
January 2021

Acknowledgments

Torben A. Lenau thanks the many students of Danmarks Tekniske Universitet (DTU) who took his course 41084 Biologically Inspired Design over the years for providing the empirical experience and contextual setting that stimulated the development of methodological support tools. He is also highly grateful for insightful discussions with and support from his wife Ingrid.

Akhlesh Lakhtakia is grateful to the Trustees of The Pennsylvania State University for a sabbatical leave of absence, the Otto Mønsted Foundation for partial financial support, and the Department of Mechanical Engineering, DTU for gracious hospitality in Fall 2019 semester. He also thanks Mercedes for wonderful spousal support during that period.

Both of us are grateful to Daniela C. A. Pigosso and Tim C. McAloone for discussions on grafting *biologically inspired design* onto *design for environment*. We thank Patrick D. McAtee for several suggestions as well as for alerting us to several errors in a draft manuscript, and the staff of Morgan & Claypool for splendid cooperation in producing this book.

Torben A. Lenau and Akhlesh Lakhtakia
January 2021

CHAPTER 1

Definitions

> "Begin at the beginning," the King said, very gravely,
> "and go on till you come to the end: then stop."
> LEWIS CARROLL, *Alice in Wonderland* (1865)

First things first, we must begin with definitions. This is all the more necessary for a rapidly emerging area such as ENGINEERED BIOMIMICRY, which encompasses both basic research on outcomes and mechanisms of diverse phenomena displayed by living organisms and the application of fundamental principles uncovered by that basic research to devise useful processes and products [1]. Engineered biomimicry can thrive in an INDUSTRIALISM, which is a society replete with manufacturing industries for mass production of a diverse array of products.

BIOMIMICRY lies within the ambit of engineered biomimicry. Although the two terms are often used as synonyms of each other, biomimicry additionally incorporates the attributes of SUSTAINABILITY evinced by the bioworld. Sustainability is defined as the maintenance of natural resources for ecological balance; hence, present-day needs are satisfied without endangering the ability of future generations to do the same [2]. Sustainability mandates the formation of those industrial ecosystems that are founded on the principles of CIRCULAR ECONOMY. The main outputs, byproducts, and wastes of every segment of a circular economy become inputs to one or more of the other segments of that economy, thereby minimizing the overall resource inputs to the circular economy [3]. The inter-relationships of engineered biomimicry, biomimicry, sustainability, and industrialism are schematically depicted in Fig. 1.1.

Design and manufacture are the two main engineering activities in any industry. Accordingly, engineered biomimicry encompasses both biologically inspired design and manufacture, as depicted in Fig. 1.2. The scope of BIOLOGICALLY INSPIRED DESIGN is the formulation of design strategies to reproduce desirable outcomes, mechanisms, and structures from the bioworld. A manufacturing action may or may not be provenanced in the bioworld.

The history of *Homo sapiens* is marked by numerous approaches to the solution of engineering problems based on solutions from the bioworld. These approaches of engineered biomimicry can be classified as bioinspiration, biomimetics, and bioreplication, as shown also in Fig. 1.2.

The goal in BIOINSPIRATION is to reproduce a biological outcome without reproducing the underlying physical mechanism(s) and the biological structure(s). As an example, powered flying machines were inspired by birds in self-powered flight. But airplanes do not flap their wings like birds, and the tails of birds are horizontal unlike the vertical tails of aeroplanes. Rotorcraft do

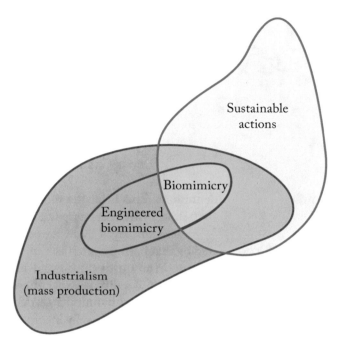

Figure 1.1: Engineered biomimicry and biomimicry within the contexts of sustainable actions and mass production.

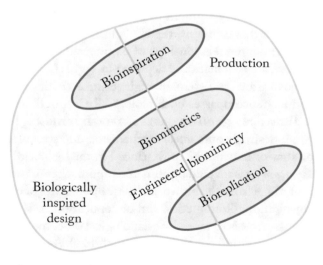

Figure 1.2: Conceptual anatomy of engineered biomimicry.

not fly like birds either. But these engineered structures do reproduce the natural outcome of moving from one location to another without being in physical contact with the ground.

BIOMIMETICS is the reproduction of a physical mechanism responsible for a specific functionality exhibited by a biological structure. The classic example of biomimetics is Velcro™ comprising dense assemblies of hooks and loops, the former emulating the hooked barbs on a burdock seed and the latter the fur of a furry animal. When a furry animal brushes against a burdock seed, the hooks get fastened to the fur. The International Standards Origanization (ISO) has formulated a set of criteria for whether a product can be considered as biomimetic [4]. The criteria relate to the biomimetic design process which was applied to develop the product and require that

(i) a function analysis has been performed on an available biological system,

(ii) the essential mechanisms in that biological system have been abstracted into a model, and

(iii) the model has been transferred and applied to design the product.

BIOREPLICATION is the direct replication of a structure found in a biological organism in order to reproduce one or more functionalities exhibited by the biological structure copied. Decoys created by nanoscale replication of the actual elytra of a female emerald ash borer for the purpose of sexually attracting male emerald ash borers provide an example of bioreplication [5].

The term BIOMATERIAL refers to either a material harvested from a biological organism to be used for the same purpose as in the organism or an artificial material used for a biological purpose. In the latter case, the term BIOCOMPATIBLE MATERIAL is also used. Bovine milk is a biomaterial of the first kind, being produced in the bioworld as a nutrient for calves and also used by humans as food. Prostheses for human hips and knees are made of biomaterials of the second kind.

BIOMANUFACTURING uses a biological process to produce a synthetic product. Thus, synthetic insulin is produced by inserting the human insulin genes in open loops of bacterial DNA to close the latter, the closed loops are inserted in bacteria which multiply rapidly in a fermentation chamber, the insulin then being harvested from the bacteria being produced in that chamber. *Escherichia coli* and *Saccharomyces cerevisiae* are commonly used species of bacteria, but yeast, a fungus, often replaces the bacteria in this biomanufacturing process [6]. A specific biomanufacturing process has at least one component that is either biomimetic or bioreplicatory.

A structure is MULTIFUNCTIONAL if it can perform two or more distinct functions that are not highly related to each other [7]. An example of multifunctionality is displayed in the bioworld by skin, which contains the organism, defines its shape and size, hosts a variety of sensors, and may be used to camouflage as well as to advertise. The fuselage of an aircraft functions both as a thermal isolator and an acoustic isolator. The well-known Swiss Army™ knife is a multifunctional tool.

The output of a MULTICONTROLLABLE structure can be controlled independently by more than one mechanisms [8]. As natural examples: the same sound can be uttered using two or three different configurations of the tongue and the buccal cavity, and multiple modes of locomotion can be used by an organism to propel itself from one location to another.

1.1   REFERENCES

[1] A. Lakhtakia and R. J. Martín-Palma (Eds.), *Engineered Biomimicry*, Elsevier, Waltham, MA, 2013. DOI: 10.1016/c2011-0-06814-x. 1

[2] M. Mulligan, *An Introduction to Sustainability: Environmental, Social and Personal Perspectives*, 2nd ed., Routledge, Abingdon, Oxford, UK, 2018. DOI: 10.4324/978131588852. 1

[3] W. R. Stahel, *Circular Economy: A User's Guide*, Routledge, Abingdon, Oxford, UK, 2019. 1

[4] ISO 18458:2015, *Biomimetics—Terminology, Concepts and Methodology*, International Standards Organization, Geneva, Switzerland, 2015. https://www.iso.org/standard/62500.html DOI: 10.3403/30274979. 3

[5] M. J. Domingue, A. Lakhtakia, D. P. Pulsifer, L. P. Hall, J. V. Badding, J. L. Bischof, R. J. Martín-Palma, Z. Imrei, G. Janik, V. C. Mastro, M. Hazen, and T. C. Baker, Bioreplicated visual features of nanofabricated buprestid beetle decoys evoke stereotypical male mating flights, *Proceedings of U.S. National Academy of Sciences*, 111:14106–14111, 2014. DOI: 10.1073/pnas.1412810111. 3

[6] N. A. Baeshen, M. N. Baeshen, A. Sheikh, R. S. Bora, M. M. M. Ahmed, H. A. I. Ramadan, K. S. Saini, and E. M. Redwan, Cell factories for insulin production, *Microbial Cell Factories*, 13:141, 2014. DOI: 10.1186/s12934-014-0141-0. 3

[7] A. Lakhtakia, From bioinspired multifunctionality to mimumes, *Bioinspired, Biomimetic and Nanobiomaterials*, 4:168–173, 2015. DOI: 10.1117/12.2258683. 3

[8] A. Lakhtakia, D. E. Wolfe, M. W. Horn, J. Mazurowski, A. Burger, and P. P. Banerjee, Bioinspired multicontrollable metasurfaces and metamaterials for terahertz applications, *Proceedings of SPIE*, 10162:101620V, 2017. DOI: 10.1117/12.2258683. 3

CHAPTER 2

What is Design?

It is not enough that we build products that function, that are understandable and usable, we also need to build products that bring joy and excitement, pleasure and fun, and, yes, beauty to people's lives.

DONALD A. NORMAN (2004)[1]

2.1 INTRODUCTION

Design has been around for as long as humans have created things. Design and making were not separate until the rise of the age of factories, since the craft-person designed the product while making it [1]. For example, a potter would make a pot by working with clay without first making drawings. This was possible as long as the product was simple and the production process was implemented close to the people using the product. However, modern products are usually very complicated and are often produced at locations far away from their users'.

This development engendered the need for more formalized DESIGN ACTIVITY whereby designers analyze user needs and create documentation so that the product can be later manufactured by others elsewhere. The documentation must be detailed and accurate in specifying form, materials, dimensions, and other variable parameters.

A design activity need not be formal and often it is not; however, it must be effective. Methods and tools are therefore developed to improve the likelihood of matching user needs to a good new product whose production is cost effective and which can be expediently disposed off after use. In writing this book, we expect that BIOLOGICALLY INSPIRED DESIGN will help designers in identifying good solution principles and even get detailed inputs for how to realize the product structure and functionality.

Apart from finding solutions to functional needs, design is also about product appearance and the messages the product sends. This is clearly obvious for clothing and automobiles, because high premiums are paid for exclusive looks. Many automobiles are designed to communicate the impression of speed and power. This is done by borrowing design features from animals with those characteristics. For example, automobile headlights are designed to remind the bystander of the eyes of tigers or lions. Cute animals inspire children's toys and sports equipment draw

---

[1]D.A. Norman, Introduction to this special section on beauty, goodness, and usability, *Human-Computer Interaction*, 19:311–318, 2004.

on visual inspiration from agile animals such as cheetahs. Biological inspiration for product appearance is a huge area, but this book is focused on how to utilize functional solutions found in the bioworld.

2.2   DESIGN THINKING

The concept of DESIGN THINKING is often invoked to distinguish design activities from scientific problem solving wherein underlying principles are uncovered systematically to find optimal solutions. In contrast, design thinking requires multiple explorations to identify a range of possible solutions from which a satisfactory one is identified.

The major difference in the thought processes of scientists and designers was exposed in an experiment more than four decades ago [2]. A group of fifth-year students from architecture and a similar group from science were asked to arrange building blocks with colored sides with the goal of maximizing the number of sides of a specific color. The results suggested that science students selected blocks in order to discover the structure of the problem, whereas architecture students generated sequences of blocks until a combination proved acceptable.

Design thinking is claimed to be suitable for solving an ill-posed problem by sketching several possible solutions to understand it from different viewpoints [1, 3]. It calls for a mindset, as can be seen from analyzing the preferred ways of working of many designers. The mindset includes a strong user focus and the will to understand the core of the problem by generating a large space of many solutions. Several of these solutions will be visualized and even prototyped before a solution is finally selected for production.

2.3   THE DESIGN OBJECT

From a first look, it seems obvious that the product is the DESIGN OBJECT. However, further analysis clarifies that the design object also includes other elements such as single components or parts within the product; the overall system within which the product functions; and the non-material services associated with its merchandizing, use, and eventual disposal.

When designing, the goal is to produce a thing to satisfy a NEED. This thing can be a physical product such as a toothbrush or an automobile, or it can be a service such as linen laundry in a hotel. Clearly, the complexity of the design process varies with the type of the product or service to be designed. The delimitation of the design object is therefore important. The toothbrush is a single component even though it is permanently assembled from a plastic handle and several clumps of brushing hairs. On the other hand, an automobile is a larger collection of single components that are configured in subsystems which together are assembled into the complete product: the automobile. However, an automobile is part of a larger system including gas stations, repair shops, roads, and parking spots, which together are necessary to provide the transportation functionality to the user. Furthermore, the product is part of a larger context which has a major impact on how the product is designed. Automobiles of different types are made to satisfy needs in diverse contexts, as exemplified by minivans to transport families with children, mobile homes for leisure activities, mobile workshops for mechanics, and taxis to transport visitors with luggage.

The BIOWORLD shows similar features. Organisms of many different types co-exist in a mutualistic relationship system that is the prerequisite for the existence of a single organism in that system. In other words, the system comprises its constituent organisms as subunits with specific roles and interfaces to the rest of the organisms. Removal of organisms of a certain type from the system can seriously alter, and even demolish, the latter. In the same way, each organism consists of several organs and other subunits with specific roles and interfaces to the rest of the organism. When seeking inspiration from the bioworld, it is therefore beneficial to also look at the larger system that the organism is part of.

One major difference between the bioworld and design activity must be noted. A new feature in an organism arises in the bioworld as a result of random modifications of parental DNA. Most of these mutations are either inconsequential or harmful, but a certain mutation may confer reproductive success in the prevailing environment. That mutation becomes more prevalent in succeeding generations. A new species emerges in consequence of a succession of numerous mutations, which makes sudden innovation impossible in the bioworld [4], the occurrence of elevated emergence rates of new species in the fossil record [5, 6] notwithstanding. For example, a marine species cannot evolve into an avian one through just one mutation. In contrast, although design activity is greatly limited by the availability of materials, tools, and expertise, disruptive innovation is possible by the interjection of a radically new concept. As an example, the emergence of the smartphones in 1992 from the predecessor telephones was a single-step achievement inasmuch as a smartphone possesses a touch screen, can email, store notes, keep a calender, and run diverse apps and widgets that would become widespread within a decade. Furthermore, smartphones began to provide very convenient access to the internet, thereby taking away a market segment from laptop manufacturers [7].

2.4   DESIGN PROCESS

While there is a general agreement that every DESIGN PROCESS starts with a user need and is expected to end with a solution, there are many models for structuring, organizing, and documenting design processes. The Pahl–Beitz model shown in Fig. 2.1 encompasses the following stages in sequence: task clarification, development of CONCEPTS (i.e., principal solutions), preliminary layout, definitive layout, and documentation [8]. Even though the model is sequential, it is recognized that many iterative loops will be made if the result of an activity is unsatisfactory.

The Cross model shown in Fig. 2.2 is organized so the different stages in a design activity form a circle [1]. The model makes it more apparent that design is an iterative activity wherein all decisions are revisited several times before a good final solution is found. Both the Pahl–Beitz model and the Cross model require a FUNCTION ANALYSIS to be undertaken before the design is specified. In the Pahl–Beitz model, function analysis takes place during concept development,

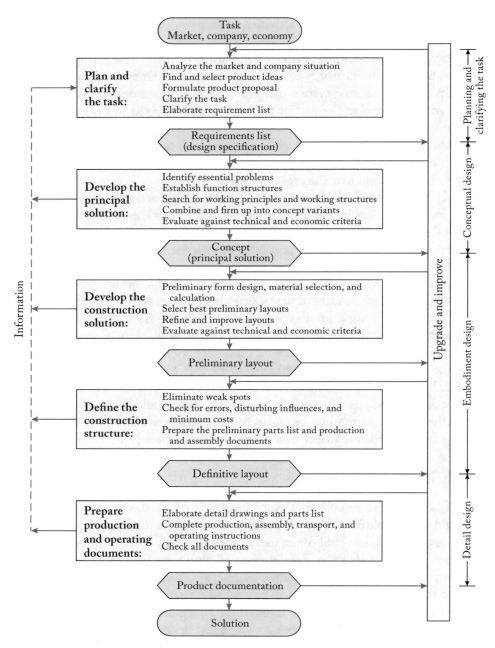

Figure 2.1: The Pahl–Beitz model of systematic design activity [8].

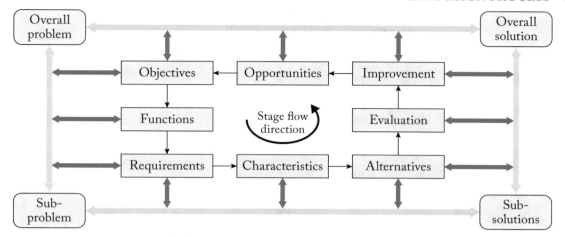

Figure 2.2: The Cross model of systematic design activity [1].

with problem identification linked to the search for working principles. In the Cross model, function analysis is used to analyze the overall design problem and break it into subproblems. A difference between the two models is the explicit focus on alternatives in the Cross model. Generating alternative solutions is an important way to figure out how the design problem is best solved. The Pahl–Beitz model, of course, recognizes this matter but the linear format of the model does not invite consideration of alternatives.

Illustrated in Fig. 2.3, the Tjalve model is a sequential model of design activity containing iterative loops [9]. With more emphasis on concept development than in the previous two models, the Tjalve model encompasses the stages of problem analysis; identification of main functions; identification of sub-functions and means; formulation of the basic structure of the product; quantification of the product structure; delimitation of materials, dimensions, and surfaces; and the overall form of the product. Whereas the Pahl–Beitz and Cross models are well suited for managing and coordinating design processes, the Tjalve model aims at guiding the designer in creative activities.

Indeed, the Tjalve model defines the product in terms of five basic attributes. These are the structure of the product with its constituent elements and relations, along with the form, material, dimensions, and surface of each element. Thus, this model is a journey from an abstract description of the product using functions toward gradually more and more concrete descriptions of the overall structure and the constituent elements. Solutions are identified for each function, the arrangement of the solutions being called the basic structure. The basic structure is typically described using symbolic graphs rather than drawings illustrating the appearance of the product. The quantified structure developed thereafter contains dimensions as well as the physical arrangement of the constituent elements.

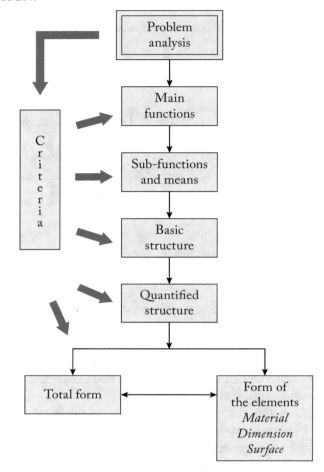

Figure 2.3: The Tjalve model of systematic design activity [9].

The Tjalve model emphasizes the need for alternatives and gives detailed inspiration for how to systematically explore different basic structures that will satisfy the functional requirements. Several alternatives are also generated for the quantified structure, wherein the constituent elements can be configured differently in relation to each other. A useful division between the total form of the product and the forms of the constituent elements allows for the search for partial solutions for single functions which later are combined into the total solution.

Finally, the integrated-product-development model shown in Fig. 2.4 emphasizes that product design is not done in isolation but in parallel and close collaboration with market- and production-oriented activities [10]. While designers consider the type of product to design, the marketing team investigates competing products and determines whether there is room for a new

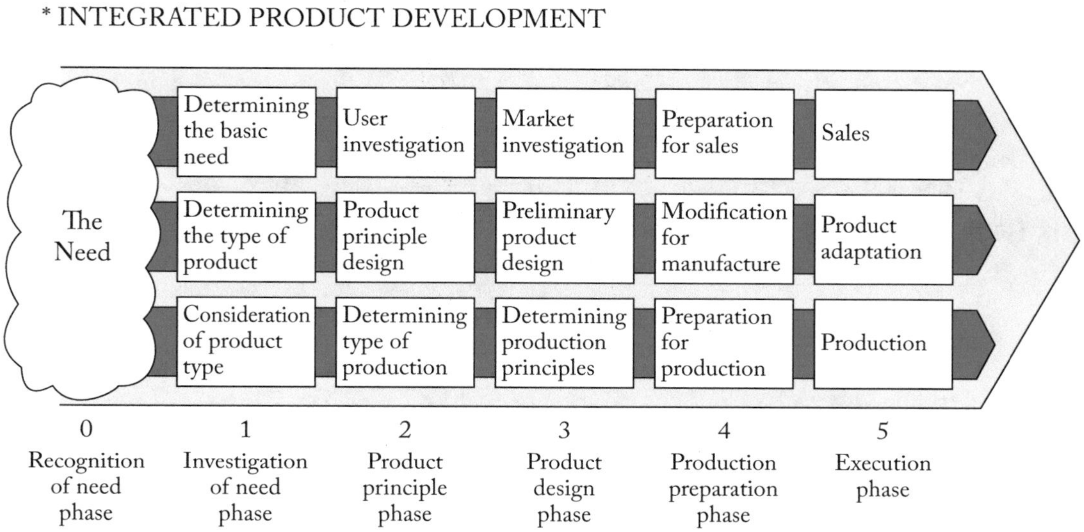

* INTEGRATED PRODUCT DEVELOPMENT

	0	1	2	3	4	5
	Recognition of need phase	Investigation of need phase	Product principle phase	Product design phase	Production preparation phase	Execution phase

Figure 2.4: The integrated-product-development model of systematic design activity [10].

product in the market, and the production team investigates diverse options for manufacturing the product.

2.4.1   TASK CLARIFICATION

A design process can be initiated by many different triggers [11]. A common trigger is the user need for a good solution, inadequate and barely adequate solutions being unsatisfactory. Another trigger is the introduction of a new technology, as exemplified by the emergence of social media triggered by the introduction of smartphones. Yet another trigger is the economic affordability of a technology. For instance, the low prices of efficient batteries have caused a boom in the use of electrical scooters and bicycles.

Once a user need has been identified, the design process has to be articulated. This is most often done either verbally or in writing, but a powerful alternative way is to provide diagrams. In order not to restrict the design process, the focus should be on highlighting the need but not on describing how it has been solved previously. This can be done by describing the consequences of satisfying the need through before/after pictures like the ones commonly used in advertisements for diet pills and other weight-reduction regimen. By visualizing the need, the designer avoids fixation on existing solutions and becomes receptive to innovative solutions. Being readily understood, graphics are also effective in communicating with stakeholders.

Many design processes actually involve re-design of existing products. Re-design can be initiated either for improved functionality or to alleviate shortcomings experienced in existing products. A detailed analysis of how users behave in the situations calling for the use of a product

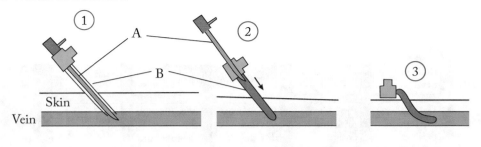

Figure 2.5: Three stages in the use of a venflon catheter for injecting a polymer tube in a vein. (1) A metal needle labeled A penetrates the skin and guides a soft polymer tube labled B into the vein. (2) The metal needle is retracted and disposed of, leaving the polymer tube in the vein. (3) The polymer tube is ready for use.

and interact with that product is typically carried out by the marketing department, but personal experiences of the designers will often lead to better results. Many companies therefore encourage their designers to directly meet users in order to understand their needs and constraints as well as how the users will actually interact with the product. User contact is also valuable for getting feedback on design proposals comprising sketches and/or prototypes.

An example of re-design is furnished by new types of the venflon catheter. A soft polymer tube in the venflon catheter is injected into a vein with the help of a stiff metal needle, as shown in Fig. 2.5. The metal needle is retracted after the polymer tube is in place and is then disposed of. Nurses revealed in interviews that the retraction as well as the disposal of the metal needle are problematic for them. Not only have two extra processes to be carried out, but the sharp needle also represents a hazard. For this reason, the venflon catheter is nowadays equipped with a small safety device which prevents the nurse from touching the needle tip after it has been retracted.

When analyzing the needs and constraints during a design activity, it can be advantageous to meet not only the direct users but also other stakeholders such as sales personnel, repair and maintenance personnel, and other persons who will come in contact with the product. Face-to-face interviews, questionnaires requiring both qualitative and quantitative answers, and personal observations provide insights. Personal experience of the product can also benefit similarly, but it also carries the risk of introducing bias. The observations and experiences of a designer are not necessarily the same as those of users. Observations are valuable since they reveal the true behavior of a user. When interviewed, users tend to give more favorable descriptions of their use patterns.

The results obtained from the analysis of user needs and existing products are described in a user-need document. This document can include information on the actual use (and misuse) of the existing products and the context of use. Sometimes, the context is explained using personas which are descriptions of typical users and their use patterns.

2.4.2 FUNCTION ANALYSIS

A way to stimulate creativity and generate new and better ways of solving problems is to formulate an abstract description of how a product functions. Instead of describing a product as an assemblage of its components, it can be described as a set of abstract functions. Another advantage is to avoid product fixation which is a risk when the names of previously used components are used. If a product is described in terms directly coupled to a specific action or form, the designer may become fixated on a specific solution and find it difficult to imagine alternatives. For example, a concrete functional description such as "drive a person from point A to point B" will fixate the designer in thinking of vehicles, whereas the abstract functional description "transport a person from point A to point B" will foster more open thinking so that a wider palette of solutions can emerge, e.g., conveyer belts and horseback riding.

A product FUNCTION is normally formulated using a verb/noun combination, such as "containing a liquid" or "cutting a piece of paper." A complete description of the functions of and within a product can be made using a functions-means tree diagram [1, 9]. The sole trapezoid at the top of the diagram describes the main function that justifies the reason why the product exists or should exist, whereas sub-functions describe what the constituent elements do. Means are physical manifestations that carry out a function. Both functions and means are explained in general terms without considering details such as shape or dimension. Figure 2.6 is a functions-means tree diagram in which the main function and sub-functions pertinent to drug delivery are identified along with the various means to accomplish each function. Note a difference in the way functions and means are described. All of the associated sub-functions are required for the realization of the main function. The more means that are recorded under a function, the more numerous are the ways of realizing that function.

A FUNCTIONAL SURFACE identifies where a specific function resides within the product [9]. A functional surface is typically marked on a sketch of the design object using hatched lines, as illustrated in Fig. 2.7. The figure shows that the two knife edges in a pair of scissors can be marked with hatched lines to indicate where the cutting function resides. Similarly, the holding function resides in a handle. The hatched lines do not invoke specific shapes, thus leaving the DESIGN ASSIGNMENT more open to innovation. A functional surface indicates the existence of an interface from the product to something else.

Another notion used to describe the functionality of a product in an abstract way is that of an ORGAN [11]. Like functional surfaces, an organ does not include information about shape and materials, but it does describe in abstract terms how a function is carried out. Two or more

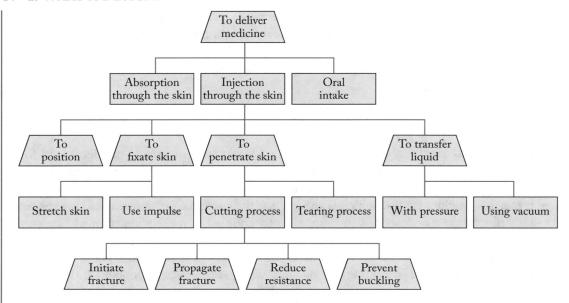

Figure 2.6: A functions-means tree diagram for ways of delivering medicine inside a patient. Each trapezoidal block contains a function, each rectangular block a means.

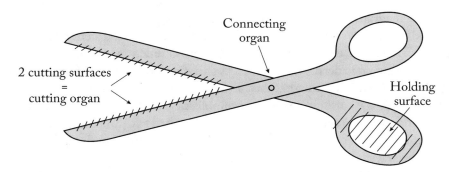

Figure 2.7: Examples of functional surfaces and organs for a pair of scissors.

functional surfaces can be combined into an organ. For example, the two cutting surfaces in a pair of scissors form a cutting organ.

Another example of an organ is the sealing organ in a container such as a bottle or a jar. A sealing organ can be realized as a lid with sealing surfaces in the lid and on the container. Using the term "lid" will automatically bring up mental pictures of existing solutions for bottles and jars. But referring to a "sealing organ" instead will make it easier to think freely of conceptually different solutions. The designer could then propose a flexible bottle where the opening is closed like a bag or the sealing organ could be a valve.

2.4.3   DESIGN BRIEF AND PRODUCT SPECIFICATION

A design assignment is typically specified using two different documents: the DESIGN BRIEF and the PRODUCT SPECIFICATION. The design brief is a visionary document that explains the context of the intended product and how future users will benefit, without going into details of the specification. The design brief is targeted toward the conceptual-design phase discussed in Section 2.4.4.

The product specification includes more detailed descriptions and is targeted toward the design phases that follow the conceptual-design phase. The product specification can be formulated in various ways. In the performance-specification approach [1], several requirements are formulated, e.g., the performance metrics that the product must satisfy and the performance characteristics that it must exhibit. For instance, the product must be manufactured in a certain range of colors and/or that it must be small enough to be stored in a standard storage space. The performance-specification approach is useful as it delivers a checklist to ensure that a DESIGN PROPOSAL is within acceptable limits. However, this approach is less helpful when comparing different design proposals and existing solutions.

The product-specification approach [10] overcomes that problem. In this approach, both requirements and criteria are stated. The requirements are fixed and must be met by the design proposal; otherwise, the design proposal is unviable. The criteria are used to compare different design proposals using evaluation matrixes (Section 2.4.5) and need to be formulated to indicate a desirable direction but not set limits. For a new toothbrush as an example, the requirements could include maximum dimensions and color; the criteria could specify that the toothbrush should be pleasant to the mouth, be easy to clean, and have a low weight.

2.4.4   CONCEPTUAL DESIGN

CONCEPTUAL DESIGN is the creative process of generating ideas for solving the overall design problem and finding partial solutions to each of the functional challenges identified in the functions-means tree diagram.

A typical way to come up with ideas is to form a brainstorming team. There are many practical approaches [1] for brainstorming, but all have in common that as large a number of ideas be generated as possible and that criticism of ideas be avoided during brainstorming sessions. Different brainstorming approaches introduce different ways of viewing the design problem; hence, using several different approaches increases the number of ideas.

Brainstorming can be done by single persons on their own, but doing it together with other people will drastically increase the chance of finding new ideas because participants will inspire each other. Also, the brainstorming session can have many different formats. One format involves a whiteboard on which a moderator writes up all the ideas that the participants propose.

In this way, a participant needs to articulate each of his/her ideas so that it is intelligible to the moderator, which also facilitates the uptake of that idea by other participants to generate additional as well as downstream ideas. Another brainstorming format lets each participant write down their ideas on colored post-it notes that are then stuck on a whiteboard. The advantage of this format is the delivery of multiple ideas by each participant, but there is a risk that the participants do not see the ideas of others and do not use those ideas to generate more ideas.

Brainstorming is a creative idea-generation method whereby the participants are not required to follow rigid rules or constraints. In contrast, systematic idea-generation methods set rigid rules for the participants, those rules possibly stimulating the participants to examine issues that they would not have otherwise thought of. Examples of systematic idea-generation methods include analysis of existing or competing solutions, TRIZ, and biologically inspired design.

Analysis of existing or competing solutions requires identification of their weaknesses and strengths so that a comparative assessment may be made of the opportunities offered by each.

TRIZ is the Russian acronym for the Theory of Inventive Problem Solving which is a method for developing innovative solutions [12]. It is based on a large (39 × 39) matrix of contradictory features (e.g., larger in volume but lighter in weight). Each field in the matrix contains a list of possible solutions developed from analyses of very large number of patents.

Biologically inspired design emerges from the thought that the bioworld offers a palette of solutions to numerous technological problems. A function analysis must be made of a candidate structure or process in the bioworld, the relevant principles must be extracted from that analysis, and finally those principles must be applied to the problem being addressed in the design activity.

It is important to differentiate between ideas and concepts. An IDEA in creative design work is basically just a principle for how to solve a problem. Application of that principle in a specific context transforms the idea into a CONCEPT that satisfies the context-specific constraints and conditions. For example, the idea of using a lid to seal a container becomes a concept in the context of closing a beer bottle, a context-specific constraint being that the lid must be able to withstand the pressure within the bottle.

Conceptual design normally starts by an examination of partial solutions to each of the functions required in the product. It is assumed that the superposition principle applies so the partial solutions can be combined to form the overall solution. One way of combining partial solutions into an overall solution is by using a morphology chart [1]. This chart is a table containing lists of possible partial solutions for each of the functions required in the product. The partial solutions can be described in words but even better with small icons for easier communication within the design team. A candidate overall solution can be formulated by selecting one partial solution for each function.

2.4.5 CONCEPT EVALUATION

The diverse concepts gathered for a given design activity can be comparatively assessed in different ways. Often, the concepts in the form of prototypes can be ranked by major stakeholders including prospective users, who can also provide feedback in the form of written and oral comments. A risk in this type of assessment is that the evaluation may be based on parameters other than what is important for the product. For example, a poorly produced prototype could be ranked poorly even though the underlying concept could lead to superior performance. To avoid such biased evaluations, the major criteria must be clearly formulated in the product specification and used for structuring the interviews with stakeholders and users.

A good tool for concept evaluation is the evaluation matrix which can be one of two types [1]:

  (i) the comparison matrix also referred to as the Pugh matrix and

  (ii) the rating matrix also called the weighted-objective-method matrix.

The comparison matrix compares each concept with a reference product which is typically an existing product. For every criterion in the product specification, each concept is rated better $(+1)$, worse (-1), or similar (0) to the reference product and the ratings are added to produce the overall rating for the concept. The major advantage of the comparison matrix is its simplicity which makes it easy to understand and discuss its results. A limitation is that all criteria need to be of equal importance since they are weighted equally.

That limitation is taken care of in the rating matrix. Each criterion is given a weight w depending on its importance. For every criterion in the product specification, each concept is given a numerical score. The overall rating for the concept is found by multiplying the score numbers with the weight for each criterion and adding the numbers together. The rating matrix should produce a better comparison of the diverse concepts than the comparison matrix. However, the rating matrix is more complicated than the comparison matrix, and stakeholders and users may find it harder to understand and accept the outcomes of a comparison.

2.4.6 TOWARD DETAILED DESIGN

What has been described so far in this chapter is normally referred to as conceptual design. The results are design concepts describing the product at a general level without going into detail about materials, shapes, and dimensions. Conceptual-design proposals are often made as hand drawings to signal that they are not finished and many details can still be altered. The more detailed design work is done thereafter: the product embodiment is carefully planned, typically using software for computer-aided design; precise dimensions are determined; tolerances are added; materials are selected; and manufacturing techniques are specified.

Besides drawings by hand and the more detailed drawings produced using computer-aided design tools, physical models are often made. Depending on their fidelity and purpose, physical models are referred to using different terms. A rudimentary model using rough materials such as cardboard, clay, and paper is typically called a mock-up. It serves to demonstrate only a few relevant aspects of the product such as physical size and how it interfaces to other products. When a physical model appears close to the final product, it is referred to as a visual model. Another type of physical model is a functional model, which only serves to demonstrate that a given principle and embodiment will actually fulfill the requirements. Finally, prototypes are traditionally used to denote models that comes very close to the final product but are typically made as one-offs. However, the terminology is drifting and many designers use the term prototype as a synonym for a mock-up or a functional model.

While all of these more detailed design activities are important elements in the overall design process, we will not go into more detail with them in this book. Biologically inspired design is typically incorporated in the conceptual-design phase.

2.5   REFERENCES

[1] N. Cross, *Engineering Design Methods—Strategies for Product Design*, Wiley, Chichester, UK, 2008. 5, 6, 7, 9, 13, 15, 16, 17

[2] B. R. Lawson, Cognitive strategies in architectural design, *Ergonomics*, 22:59–68, 1979. DOI: 10.1080/00140137908924589. 6

[3] N. Cross, *Design Thinking: Understanding how Designers Think and Work*, Berg, Oxford, UK, 2011. DOI: 10.5040/9781474293884. 6

[4] D. Adriaens, Evomimetics: The biomimetic design thinking 2.0, *Proceedings of SPIE*, 10965:1096509, 2019. DOI: 10.1117/12.2514049. 7

[5] N. Eldredge and S. J. Gould, Punctuated equilibria: An alternative to phyletic gradualism, *Models in Paleobiology*, T. J. M. Schopf, Ed., pages 82–115, Freeman Cooper, San Francisco, CA, 1972. 7

[6] M. J. Benton and P. N. Pearson, Speciation in the fossil record, *Trends in Ecology and Evolution*, 16:405–411, 2001. DOI: 10.1016/s0169-5347(01)02149-8. 7

[7] C. M. Christensen, M. Raynor, and R. McDonald, What is disruptive innovation?, *Harvard Business Review*, 93(12):44–53, 2015. 7

[8] G. Pahl, W. Beitz, J. Feldhusen, and K.-H. Grote, *Engineering Design: A Systematic Approach*, 3rd ed., Springer, London, UK, 2007. DOI: 10.1007/978-1-84628-319-2. 7, 8

[9] E. Tjalve, *A Short Course in Industrial Design*, Butterworth, London, UK, 1979. DOI: 10.1016/C2013-0-00824-9. 9, 10, 13

[10] M. M. Andreasen and L. Hein, *Integrated Product Development*, IFS (Publications) Ltd., Kempston, UK, 1987. 10, 11, 15

[11] M. M. Andreasen, C. T. Hansen, and P. Cash, *Conceptual Design: Interpretations, Mindset and Models*, Springer, Cham, Switzerland, 2015. DOI: 10.1007/978-3-319-19839-2. 11, 13

[12] J. F. V. Vincent, O. A. Bogatyreva, N. R. Bogatyrev, A. Bowyer, and A.-K. Pahl, Biomimetics: Its practice and theory, *Journal of the Royal Society Interface*, 3:471–482, 2006. 16

C H A P T E R   3

Engineered Biomimicry: Solutions from the Bioworld

> If a group of engineers, mindful of our need to tap natural energy sources, were to embark on designing a machine that would pump water out of the ground over an area of 100 square meters continuously, and would boil off the water into steam, using only the energy directly from the sun for the whole process, it is possible that they might do it. But their finished machine would certainly never resemble a tree!
>
> ERIC R. LAITHWAITE (1988)[1]

Although we humans have long been envious of feats of performance displayed by a variety of animal species [1], and we have been creative in emulating and even surpassing some of those feats, biomimicry began to acquire an organizational framework only during the 1990s. Coinage of the term BIOMIMETICS is usually attributed to Otto Schmitt during the late 1950s [2]. The similar term BIOMIMESIS coined during the next decade [3] does not have much currency nowadays. The term BIONICS, once synonymous with biomimetics [4], is nowadays employed in English exclusively to the science and practice of replacing an organ in a living being by a prosthesis. The umbrella term BIOMIMICRY has come to subsume its precedents, although one (namely, bionics) survives as BIONIK in German.

Biomimicry opens "the possibility of a new industrialism that is more attuned to nature's needs" [5] and therefore intersects with the discipline of SUSTAINABLE DESIGN. As discussed in Chapter 1, ENGINEERED BIOMIMICRY does not require consideration of sustainability. In this chapter, we first lay out the case for engineered biomimicry, then present a few representative examples, identify some characteristics of the solutions available in the bioworld for technological problems, and finally discuss the importance of having biologists on design teams for bioworld solutions.

3.1 THE CASE FOR ENGINEERED BIOMIMICRY

Charles Darwin used the word EVOLVE only once in the first edition [6] and just 16 times in the sixth edition [7] of his most famous book *On The Origin of Species*. Instead, he used the term

---

[1]E. R. Laithwaite, Gaze in wonder: an engineer looks at biology, *Speculations in Science and Technology*, 11:341–345, 1988.

DESCENT WITH MODIFICATION to describe the origin of new species. Most traits of a child are derived from those of its parents, but some modifications may occur.

Later scientists realized that genes are the vehicles for heritability or descent and that imperfect replication of parental DNA results in random modifications called mutations. Most mutations are either inconsequential or harmful, but a certain mutation may confer reproductive success in the prevailing environment. That mutation becomes more prevalent in succeeding generations, the process being called NATURAL SELECTION.

Whereas mutations are random, natural selection is not. Only those mutations that lead to better adaptation to altering or altered environments are successful. A continuum of morphological varieties thus arises in a species. A series of successful mutations, genetic transfer from one population to another as a result of migration, and random changes in the frequencies of certain genes are mechanisms which eventually result in a new species that does not have morphological intermediates between it and the older species.

As of now, about 1.3 million species have been identified, but some 86% of terrestrial species and 91% of marine species are estimated to still await description [8]. Add the 4 billion species that are estimated to have gone extinct [9] since life began on our planet some 4 billion years ago [10]. Each of those species can be considered as being successful for a certain period, dying out only when the environmental conditions were no longer conducive enough to sustain it.

The success of any mutation cannot be predicted and there is no prescient agency for natural selection. Still, looking at the history of the bioworld, both recent and in the prehistoric past, we may regard all species as data points in a multidimensional space. The mutually orthogonal axes of this space are physical variables (such as ambient temperature, ambient pressure, and mass density) and performance characteristics (such as speed of locomotion, longevity, and fecundity). Each species as a data point represents a successful experiment.

Since the laws of physics hold sway over every biological process just as completely as over every technological operation, we should then consider the bioworld as a repository of answers to billions of technological questions [11]. Some of those answers may not be optimal for our technological requirements but can still illuminate possible research directions. Other answers may be used by us without much fuss. Furthermore, the bioworld contains a plethora of processes some of whom can be replicated either partially or wholly in industrial operations. No wonder, humans have long been inspired by attractive outcomes and functionalities evident in plants and animals.

3.2   ENGINEERED BIOMIMICRY

Engineered biomimicry encompasses both basic research on outcomes and mechanisms of diverse phenomena displayed by living organisms and the application of fundamental principles uncovered by that basic research to devise useful processes and products. Engineered biomimicry is classified into bioinspiration, biomimetics, and bioreplication, as shown in Fig. 3.1 [12], based

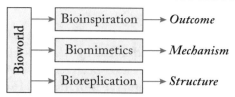

Figure 3.1: Classification of engineered biomimicry into bioinspiration, biomimetics, and bioreplication.

on whether outcomes, mechanisms, or structures in the bioworld are aimed for reproduction in technoscientific settings.

3.2.1   BIOINSPIRATION

Ancient stories provide numerous examples of the human desire to fly. After rescuing two children from a sacrificial altar, a flying ram became the constellation Aries in Greek mythology. Zeus, the king of Greek gods, had a winged steed named Pegasus. Quetzalcoatl, the Aztec god of wind and learning, was a winged serpent. Hindu mythology is replete with flying chariots and palaces. Mohammad, the prophet of Islam, was flown to heaven by a white mule-donkey hybrid named Burāq.

Some 500 years ago, Leonardo Da Vinci (1452–1519) studied birds to conceptualize several flying contraptions which evidently never took off. Sir George Cayley (1773–1857) made a pilotless glider that did fly in 1804. Orville and Wilbur Wright were to first to successfully fly a heavier-than-air machine with a person onboard, on December 17, 1903. The emergence of aeroplanes inspired by birds in self-powered flight is an excellent example of bioinspiration, but birds and aeroplanes have different flying mechanisms. The goal in bioinspiration is to reproduce a biological outcome but not the underlying biological mechanism(s) and structure(s).

3.2.2   BIOMIMETICS

Biomimetics is the reproduction of a physical mechanism responsible for a specific outcome or functionality exhibited by a biological structure. Greek mythology furnishes the classical example of biomimetics through Icarus, a flying human who escaped from a Cretan prison using wings made of feathers and wax. Sadly, he perished after the wax melted when he flew too close to the sun.

A modern example is that of insulin, a hormone produced naturally in mammalian pancreas but nowadays modified and synthesized in either yeasts or *Escherichia coli* bacteria [13, 14]. Yet another example of biomimetics is Velcro™ that comprises dense assemblies of hooks and loops, the former emulating the hooked barbs on a burdock seed and the latter, the fur of a furry animal. The commercialization of this biomimetic analog of a natural mechanism of adhesion is a fascinating story of determination [15].

3.2.3 BIOREPLICATION

Bioreplication is the direct replication of a structure found in a biological organism in order to reproduce one or more functionalities exhibited by the biological structure copied. During the last ten years, diverse physical techniques have been harnessed to replicate several biological structures such as the eyes and wings of several types of insects [16]. The techniques include the sol-gel method; atomic layer deposition; physical vapor deposition; and some combination of imprint lithography, casting, and stamping [17]. Some of these techniques are more suitable for reproducing surface features, others for bulk three-dimensional structures.

3.3 EXAMPLES OF ENGINEERED BIOMIMICRY

3.3.1 BIOINSPIRED COMPUTATIONAL TECHNIQUES

Every multicellular organism contains one or more networks in which information is sensed, transmitted, processed, transmitted again, and then acted upon. Relying on physical and chemical phenomena, all of these processes are quantitative and therefore may be mathematically modeled by us, albeit not always easily.

Mathematical models of many biological processes employ differential equations to relate spatial and temporal gradients of physical quantities, such as the concentrations of some chemicals, partial pressure of various fluids, and the electric charge density transported by ions. Initial and boundary conditions therefore must be concurrently considered [18, 19]. Successful examples include models of oxygen-deficient dermal wounds [20] and cancer growth [21].

Often, the data gathered about a biological process is both discrete and huge, as exemplified by tumor growths [22] and neuronal activity [23]. To analyze these data, mathematical methods commonly used for time series [24] and dynamical systems [25] are pressed into service.

The two foregoing paragraphs provide examples of mathematical methods applied to understand biological processes. Are some mathematical methods to analyze non-biological phenomena inspired by the bioworld? An affirmative answer to that question has emerged in modern times [26]. Inspired by the structure of animal brains, ARTIFICIAL NEURAL NETWORKS (ANNs) are being used for pattern recognition tasks, including speech recognition, machine translation, video games, and traffic control; FUZZY LOGIC seeks to emulate human cognition for automated decision making; SWARM INTELLIGENCE guides mathematical investigations of emergent phenomena; GENETIC ALGORITHMS are often used for optimization; and so on. Let us focus on two of these bioinspired computational techniques.

Artificial Neural Networks

ANNs have been inspired by animal brains which are networks of neurons connected to other neurons through synapses [27, 28]. In an ANN, neurons are replaced by nodes and synapses by connections, as depicted schematically in the top panel of Fig. 3.2.

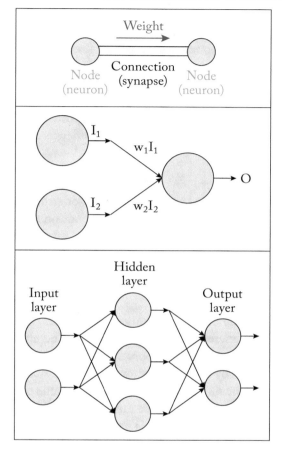

Figure 3.2: Top: Schematics for artificial neural networks.

The middle panel of Fig. 3.2 shows two nodes providing inputs I_1 and I_2 to a node whose output is denoted by O. The output is related to the inputs by a nonlinear function $f(x)$ such that

$$O = \begin{cases} 0, & w_1 I_1 + w_2 I_2 < b, \\ f(w_1 I_1 + w_2 I_2), & w_1 I_1 + w_2 I_2 \geq b. \end{cases} \tag{3.1}$$

The on/off characteristic of real neurons is simulated by the conditionality on the right side of Eq. (3.1), with b as the bias or the threshold value of the argument x of $f(x)$, and the relative importance of the inputs coded through the weights w_1 and w_2.

An ANN can have several input nodes arranged in a layer and several output nodes arranged in a different layer. In between is at least one layer of hidden nodes, called thus because these nodes have no direct connection to: (i) the sensors providing data to the input layer and

(ii) the actuators implementing actions controlled by the output layer. The bottom panel of Fig. 3.2 shows an ANN in which information moves in the forward direction, i.e., from the input nodes, through the hidden nodes, to the output nodes. ANNs of other types can have backward connections and even loops.

Known sets of input-output data are used to train an ANN, i.e., determine the weights. More training data will determine the weights better (usually but not always), the assumption being that the ANN learns just like a biological brain. After the ANN is deemed to have learned enough, it can be fed data to predict the output with confidence.

Genetic Algorithms

Genetic algorithms are commonly used to design a device or structure to meet a numerical criterion for performance [29]. The device performance depends on the values of a certain number (say N) of characteristic variables. The algorithm begins by randomly selecting $M_1 > 1$ sets of the N characteristic variables. A performance function denoted by p is calculated for every one of the M_1 sets. If $p \geq b_1$ for a specific set, where b_1 is a threshold value, then that particular set is retained; if not, that set is eliminated. The result is that $\bar{M}_1 \leq M_1$ sets survive to reproduce the next generation comprising M_2 new sets.

The simplest reproduction method is mutation, whereby each new set is based on a single surviving set of the previous generation. If the population is being doubled by mutation (i.e., $M_2 = 2\bar{M}_1$), each set of the old generation is reproduced twice, once as itself and once by multiplying its characteristic variables by a random factor. A more complex method of reproduction is crossover, whereby each set of the new generation is based on some combination of the surviving sets of the previous generation. The performance function p is calculated for each one of the M_2 sets. If $p \geq b_2$ for a specific set, where $b_2 > b_1$ is a new threshold value, then that particular set is retained; if not, that set is eliminated.

This process of creating new generations continues until a criterion for terminating it is satisfied. At that stage, several devices satisfying the performance criterion $p \geq \max\{b_1, b_2, \ldots\}$ could have been identified. Then comes the task of selecting and making at least one of those devices.

3.3.2   BIOMIMETIC PRODUCTION OF HUMAN INSULIN

The peptide hormone insulin began to be used in 1922 to treat diabetic patients. In a normal person, insulin is produced in the pancreas where it is stored well in excess of daily needs. A series of biochemical reactions in response to elevated concentration of glucose in blood triggers the release of insulin from the pancreas. Its half-life ranging between four and six minutes, it lasts outside the pancreas for about an hour, and is eventually cleared by the liver and the kidneys.

The human insulin molecule has 51 amino acids, its molecular formula being $C_{257}H_{383}N_{65}O_{77}S_6$. Insulin is produced and stored in the pancreas as a hexamer, i.e., an aggregate of six molecules. The hexamer is very stable. Insulin is released from the pancreas as a monomer, which acts very rapidly.

Until about three decades ago, virtually all insulin injected into patients was derived from the glands of either cows or pigs obtained as waste products from abattoirs. Bovine insulin differs from human insulin in only three amino acids, porcine insulin in just one. Fourteen cattle or 70 pigs had to be slaughtered to harvest enough insulin to last a patient for a year. However, the responses of some patients were unpredictable and some patients had severe reactions.

Research began in the 1970s for a biomimetic route to synthesize human insulin itself [13]. That research has been wildly successful [14]. The sequence of biochemical reactions in mammalian pancreas is replicated in yeasts and bacteria. The reproduction of yeasts and bacteria can be regulated fairly easily, which then eliminates the need for continually harvesting mammalian pancreas. Moreover, as the production process is initiated with human insulin, the biomanufactured insulin is maximally compatible with human patients.

Pancreatic Production

The production of a molecule called preproinsulin is encoded in a gene found in chromosome 11 in the nuclei of human cells. A chromosome is a DNA molecule comprising nucleotides of four different types arranged into two strands that are coupled to each other by hydrogen-hydrogen bonds. There are also packing proteins in the chromosome to keep the DNA molecule untangled.

Every nucleotide contains a nitrogenous base. There are four types of nitrogenous bases: adenine, thymine, guanine, and cytosine. Whereas adenine can form a hydrogen-hydrogen bond only with thymine, guanine can form a hydrogen-hydrogen bond only with cytosine. Thus, adenine and thymine are mutually complementary, and so are guanine and cytosine. The sequence of bases in one strand of a DNA molecule is matched by the sequence of complementary bases on the accompanying strand.

Three consecutive bases form a codon. A codon contains the instructions to produce a protein-creating amino acid. There are 22 protein-creating amino acids. Of the 64 codons possible, 61 provide instructions for producing 20 of those amino acids. Some amino acids can be produced by more than one codon. The final two protein-creating amino acids are synthesized through complex reactions.

A short sequence of amino acids is called a peptide. A long sequence of amino acids is called a polypeptide or a protein. Three codons are used to indicate the end of an amino-acid sequence, the start of that sequence being signaled in a more complex way.

Thus, the DNA molecule in a chromosome comprises two complementary chains of codons. A gene is a sequence of codons that contains instructions to produce a molecule that performs a function. Some genes contain instructions to produce proteins, others to produce different types of RNA molecules. An RNA molecule is a single strand of nucleotides of four types, each containing either adenine, thymine, guanine, or uracil (different from cytosine found in DNA molecules).

The DNA molecule can then be considered as two chains of identical genes, but it also contains codon sequences that may either have no purpose or whose purpose has yet not been discovered.

The process of insulin production in pancreatic cells begins when an enzyme called RNA polymerase, accompanied by molecules called transcription factors, attaches to a region in the DNA molecule just before the start of the preproinsulin-producing gene. Then the two DNA strands separate, and RNA nucleotides attach via hydrogen-hydrogen bonds to the nucleotides in one of the two strands of the DNA molecule until the stop codon is encountered. At that stage, the RNA molecule dissociates from the DNA strand, and the two strands of the DNA molecule couple again.

The RNA molecule thus synthesized is called a messenger RNA (mRNA). It has the instructions to produce preproinsulin. That process begins when a transfer RNA (tRNA) molecule and a ribosome attach themselves to the start codon of the mRNA molecule. Depending on the next codon, the appropriate amino acid attaches itself to the end of the tRNA molecule. The ribosome then translocates to the next codon, and the next appropriate amino acid attaches itself to the previous amino acid. This elongation of the tRNA molecule continues until the stop codon is reached. At that stage, the single-chain preproinsulin molecule is attached to the original tRNA molecule. The two then dissociate.

A chemical reaction in the endoplasmic reticulum in the pancreas causes the removal of 12 amino acids from the preproinsulin molecule, which then folds into two linear chains connected by a peptide. The resulting molecule is called proinsulin. Removal of the connecting peptide turns the proinsulin molecule into the insulin molecule.

Biomimetic Production of Insulin

This complex process had to be reproduced biomimetically. Researchers chose *E. coli*, a bacterium that contains a circular chromosome [13]. Some strains of *E. coli* also contain a circular plasmid, which is a genetic structure that is not a chromosome. The gene *INS* is responsible for producing preproinsulin in humans. This gene is inserted in the plasmids of some bacteria, as shown schematically in Fig. 3.3. As the bacteria with the altered plasmids reproduce in a fermentation chamber, the number of the altered plasmids increases. Biochemical reactions are then used to harvest proinsulin molecules, which are then converted chemically to insulin molecules. Some manufacturers use yeasts in place of *E. coli*.

The dominant mode of reproduction in both types of single-celled organisms is asexual. In a process named mitosis, a cell elongates and then divides once to form two identical cells. Both of these cells are genetically identical to the cell that underwent mitosis.

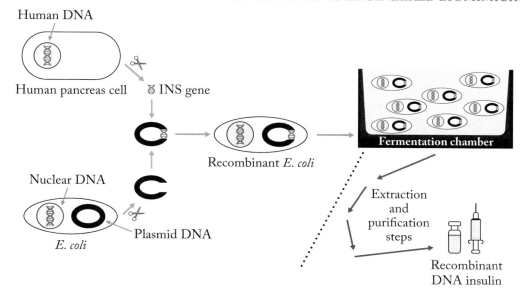

Figure 3.3: Schematic for biomimetic production of insulin.

The entire biomimetic process is initiated by some copies of *INS*, but no more are needed after production begins. The proclivity of single-cell organisms to reproduce rapidly *via* mitosis makes the biomimetic production of insulin economically viable.

Fast-acting insulins are produced by slight interchanges of codons in the initiating copies of the human genes to minimize the tendency to form hexamers. The type of interchange selected regulates the ratio of monomers to hexamers. Intermediate-acting insulins are produced by adding chemicals that help maintain hexamers. Long-acting insulins are produced by slight modifications of an amino acid. Thus, a therapeutically significant functionality is imparted to biomanufactured insulin in comparison to insulin produced in the pancreas.

3.3.3 BIOREPLICATED VISUAL DECOYS OF INSECTS

An industrially scalable bioreplication process with nanoscale fidelity has been devised to produce visual decoys of females of the buprestid insect species *Agrilus planipennis*, commonly called the emerald ash borer (EAB). The decoys are more successful than freshly sacrificed females in luring males of the species toward attempted copulation followed by electrocution [30, 31], thereby providing forestry managers a tool to limit the spread of the invasive species.

The emerald ash borer is a native of northeast Asia. Its shipborne arrival in North America was detected in 2002. That very year, it was identified as devastating ash trees. EAB females deposit eggs in the bark of ash trees; the EAB larvae chew long meandering tunnels in the

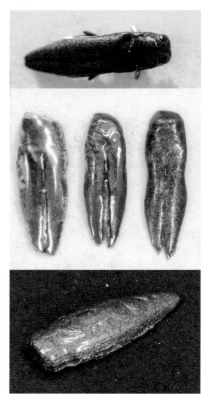

Figure 3.4: Top: Female of the species *Agrilus planipennis*. Middle: Three types of bioreplicated decoys produced with an industrially scalable process [30]. Bottom: 3D-printed decoy [35].

trunks as they feed, thereby disrupting the transport of nutrients and water to the leaves; and adults chew their way back to the bark and exit the trunk [32]. EAB are thriving in North America in the absence of natural predators and parasitoids. Although their populations spread about 20 km per year, long-distance transport of wood products allows them to colonize far-flung areas. Ash wood being used for numerous purposes, the destruction of ash trees is having a severe economic impact. Furthermore, as other invasive species move into the affected areas, native species suffer from habitat reduction and the soil chemistry changes [32].

EAB do not have sex pheromones to attract mates, relying instead on visual communication. Adult EAB are conspicuous by their bright metallic green elytra (hardened forewings), as shown in the top panel of Fig. 3.4. Adult EAB males patrol tree canopies for adult EAB females resting and feeding on ash leaves. After seeing a female from as high as 100 cm, a male drops like a paratrooper toward her and makes vigorous attempts to copulate [33].

A visual decoy looking very similar to an EAB female with its elytra folded over its body would be necessary to lure EAB males. The decoy's color must be iridescent green to contrast against the the background of ash foliage. Additionally, ~10-μm surface features present on the elytra must be reproduced on the decoy.

An industrially scalable bioreplication process was therefore devised [30]. This process involved two major stages. In the first stage, a pair of matching positive epoxy and negative nickel dies were bioreplicated from an euthanized female EAB. The negative die was made by the deposition of a ~500-nm-thick conformal film of nickel on the upper surface of the euthanized female EAB in a low-pressure chamber. The nickel thin film was then thickened by electroforming to about 100 μm. The female EAB was then plucked out, leaving behind a negative die with fine-scale features, the conformal film comprising ~22-nm-diameter nickel grains. A positive die of epoxy was made from the negative die of nickel using several casting steps and the deposition of a conformal thin film of chalcogenide glass.

In the second stage, a sheet of poly(ethylene terephthalate) (PET) was hot stamped between the pair of matching dies. The PET sheet had been previously coated on the upper side with a quarter-wave-stack Bragg filter [34] made of two distinct polymers to reflect normally incident green light and on the lower side by black paint to absorb visible light of all other colors. Light stamping between the pair of matching dies kept the Bragg filter intact. However, heavy stamping for better reproduction of the fine-scale features of the elytra pulverized the Bragg filter, for which reason the lower side of the decoy was spray-painted metallic green, again to mimic the actual color of the EAB elytra. The middle panel of Fig. 3.4 is a photograph of bioreplicated decoys of three different types.

In a preliminary field experiment, males of the related species *A. bigutattus* were targeted, the inter-species attraction having been previously recorded by entomologists. The bioreplicated decoys were 40% more effective in luring males than dead EAB females [30]. The lower effectiveness of the dead EAB females is indicative of the suboptimality of many biological phenomena, as discussed in Section 4.5.

The effectiveness of the bioreplicated decoys was evaluated against that of 3D-printed decoys, an example of which is shown in the bottom panel of Fig. 3.4. Although EAB males were almost equally attracted to decoys of both types, they would fly toward and alight on the bioreplicated decoys for a couple of seconds, but they would break off midway toward the 3D-printed decoys and veer away. The absence of the ~10-μm surface features on the 3D-printed decoys rendered them insufficiently authentic on closer inspection by the EAB males.

In the third field experiment [31], the bioreplicated decoys were offered to EAB males. These decoys evoked complete attraction, paratrooper flight, and attempted copulation from EAB males. Some decoys were electrically wired for alighting males to be electrocuted. The electrocuting decoys could assist forestry managers in slowing the spread of the pest species.

The bioreplication process for industrial-scale production of these decoys was sped up [36] by making the negative nickel die from an array of several female EABs instead of only one.

Also, the positive die was eliminated by a decision to fill up the multiple cavities of the negative die with the thermally curable liquid polymer poly(dimethyl siloxane). Multiple decoys made simultaneously were painted metallic green.

The tale of EAB decoys is one in which a biological structure is directly replicated by technoscientists in order to fulfill a societal goal: to eliminate a pest species, or at least reduce its proliferation. Can this nanoscale bioreplication process also assist biologists in answering certain questions that cannot be answered otherwise? The answer is a guarded "yes." For instance, the spectral ranges of buprestid vision systems could be determined by coloring the decoys red, blue, or yellow, or even ultraviolet. Of course, humans cannot see ultraviolet, but many insect species can [37]. The same bioreplication technique could be applied to determine the spectral ranges of the vision systems of their predator species. Even evolutionary scenarios could be investigated by determining the aversion or affinity of a predator species to color mutations in a prey species.

3.4 DESIGN TEAMS FOR BIOWORLD SOLUTIONS

The examples of engineered biomimicry presented in some detail in this chapter strongly indicate that this topic transcends the boundary between science and engineering. Until perhaps the middle of the 19th century, there was no distinction between engineers and scientists. The explanation of natural phenomena, today considered the domain of scientists, and the commercially viable exploitation of those phenomena for the betterment of the human condition, today considered the domain of engineers, were conjoint goals of a person who functioned either as a scientist or as an engineer in different phases of professional life. Sometimes, that person even functioned concurrently as an engineer and a scientist.

The English word SCIENTIST was coined in 1834 for someone dedicated to the pursuit of new knowledge in any branch of science [38, 39]. In the ensuing decades, scientists were differentiated from engineers, the former as the discoverers of new facts in nature and formulators of potentially verifiable theories to explain those facts, the latter as those who apply scientific knowledge to solve practical problems at a cost that society can bear.

This differentiation is less pronounced nowadays, especially when multidisciplinary teams are formed to undertake complex research projects, whether at universities or in industries or in university-industry consortia. Teams comprise physicists, chemists, materials scientists, mechanical engineers, chemical engineers, electrical engineers, medical scientists, etc., as dictated by project requirements.

The scope of biologically inspired design is the formulation of design strategies to reproduce desirable outcomes, mechanisms, and structures from the bioworld. The practice of biologically inspired design requires both scientists and engineers to work collaboratively, just as for other types of complex research projects with an industrial focus. There is, however, a crucial issue.

Such a team must have biologists each of whom who specializes in a particular species that exhibits a desirable outcome, mechanism, or structures. But the expertises of these biologists may not be enough for optimal design. There may be other species—in even other genera, families, orders, and classes—that could be the sources of better designs than the species that could have inspired the formation of the biologically inspired design project. The independent evolution of a similar feature in two very different species is called HOMOPLASY, if that feature was not inherited from a recent common ancestor [40]. The evolution of a feature that has similar form or function in widely different species attests to the robustness of that feature. But not every manifestation of a certain feature would be equally efficacious for the process or product to be designed. A biologist who is focused on a desirable outcome, mechanism, or structure—instead of a particular species—could therefore guide the other team members to better choices [41].

3.5   REFERENCES

[1] H. D. Wolpert, The world's top olympians, *Engineered Biomimicry*, A. Lakhtakia and R. J. Martín-Palma, Eds., pages xix–xxv, Elsevier, Waltham, MA, 2013. DOI: 10.1016/c2011-0-06814-x. 21

[2] J. M. Harkness, A lifetime of connections: Otto Herbert Schmitt, 1913–1998, *Physics in Perspective*, 4:456–490, 2002. DOI: 10.1007/s000160200005. 21

[3] E. O. Attinger, Biomedical engineering: From biomimesis to biosynthesis, *Bioengineering: An Engineering View*, G. Bugliarello, Ed., pages 213–246, San Francisco Press, San Francisco, CA, 1968. 21

[4] O. H. Schmitt, Where are we now and where are we going? *Proceedings of USAF Air Research and Development Command Symposium on Bionics*, pages 483–486, WADD Technical Report 60-600, Dayton, OH, Sept. 1960. 21

[5] E. R. Johnson and J. Goldstein, Biomimetic futures: Life, death, and the enclosure of a more-than-human intellect, *Annals of the Association of American Geographers*, 105:387–396, 2015. DOI: 10.1080/00045608.2014.985625. 21

[6] C. Darwin, *On the Origin of Species by Means of Natural Selection*, 1st ed., John Murray, London, UK, 1859. 21

[7] C. Darwin, *On the Origin of Species by Means of Natural Selection*, 6th ed., John Murray, London, UK, 1882. 21

[8] C. Mora, D. P. Tittensor, S. Adl, A. G. B. Simpson, and B. Worm, How many species are there on earth and in the ocean? *PLoS Biology*, 9:e1001127, 2011. DOI: 10.1371/journal.pbio.1001127. 22

[9] G. G. Simpson, How many species? *Evolution*, 6:342, 1952. DOI: 10.2307/2405419. 22

[10] M. S. Dodd, D. Papineau, T. Grenne, J. F. Slack, M. Rittner, F. Pirajno, J. O'Neil, and C. T. S. Little, Evidence for early life in Earth's oldest hydrothermal vent precipitates, *Nature*, 543:60–64, 2017. DOI: 10.1038/nature21377. 22

[11] V. Davidov, Biomimicry as a meta-resource and megaproject, a literature review, *Environment and Society: Advances in Research*, 10:29–47, 2019. DOI: 10.3167/ares.2019.100103. 22

[12] A. Lakhtakia and R. J. Martín-Palma (Eds.), *Engineered Biomimicry*, Elsevier, Waltham, MA, 2013. 22

[13] J. A. Kehoe, The story of biosynthetic human insulin, *Frontiers in Bioprocesssing*, S. K. Sikdar, M. Bier, and P. W. Todd, Eds., pages 45–49, CRC Press, Boca Raton, FL, 1990. DOI: 10.1016/c2011-0-06814-x. 23, 27, 28

[14] N. A. Baeshen, M. N. Baeshen, A. Sheikh, R. S. Bora, M. M. M. Ahmed, H. A. I. Ramadan, K. S. Saini, and E. M. Redwan, Cell factories for insulin production, *Microbial Cell Factories*, 13:141, 2014. DOI: 10.1186/s12934-014-0141-0. 23, 27

[15] S. D. Strauss, *The Big Idea: How Business Innovators Get Great Ideas to Market*, pages 14–18, Dearborn Trade Publishing, Chicago, IL, 2002. 23

[16] D. P. Pulsifer and A. Lakhtakia, Background and survey of bioreplication techniques, *Bioinspiration and Biomimetics*, 6:031001, 2011. DOI: 10.1088/1748-3182/6/3/031001. 24

[17] R. J. Martín-Palma and A. Lakhtakia, *Nanotechnology: A Crash Course*, SPIE Press, Bellingham, WA, 2010. DOI: 10.1117/3.853406. 24

[18] D. S. Jones, M. J. Plank, and B. D. Sleeman, *Differential Equations and Mathematical Biology*, 2nd ed., CRC Press, Boca Raton, FL, 2009. DOI: 10.1201/9781420083583. 24

[19] W. E. Schiesser, *Differential Equation Analysis in Biomedical Science and Engineering*, Wiley, Hoboken, NJ, 2014. DOI: 10.1002/9781118705292. 24

[20] C. Xue, A. Friedman, and C. K. Sen, A mathematical model of ischemic cutaneous wounds, *Proceedings of U.S. National Academy of Sciences*, 106:16782–16787, 2009. 24

[21] J. S. Lowengrub, H. B. Frieboes, F. Jin, Y.-L. Chuang, X. Li, P. Macklin, S. M. Wise, and V. Cristini, Nonlinear modelling of cancer: Bridging the gap between cells and tumours, *Nonlinearity*, 23:R1–R91, 2010. DOI: 10.1088/0951-7715/23/1/r01. 24

[22] K. A. Rejniak and A. R. A. Anderson, Hybrid models of tumor growth, *Wiley Interdisciplinary Reviews: Systems Biology and Medicine*, 3:115–125, 2011. DOI: 10.1002/wsbm.102. 24

[23] S. J. Schiff, K. Jerger, D. H. Duong, T. Chang, M. L. Spano, and W. L. Ditto, Controlling chaos in the brain, *Nature*, 370:615–620, 1994. DOI: 10.1038/370615a0. 24

[24] W. A. Woodward, H. L. Gray, and A. C. Elliott, *Applied Time Series Analysis with R*, 2nd ed., CRC Press, Boca Raton, FL, 2017. DOI: 10.1201/b11459. 24

[25] R. J. Brown, *A Modern Introduction to Dynamical Systems*, Oxford University Press, Oxford, UK, 2018. 24

[26] W. Banzhaf, Evolutionary computation and genetic programming, *Engineered Biomimicry*, A. Lakhtakia and R. J. Martín-Palma, Eds., pages 429–447, Elsevier, Waltham, MA, 2013. DOI: 10.1016/c2011-0-06814-x. 24

[27] W. S. McCulloch and W. Pitts, A logical calculus of the ideas immanent in nervous activity, *Bulletin of Mathematical Biophysics*, 5:115–133, 1943. DOI: 10.1007/bf02478259. 24

[28] I. N. da Silva, D. H. Spatti, R. A. Flauzino, L. H. B. Liboni, and S. F. dos Reis Alves, *Artificial Neural Networks: A Practical Course*, Springer, Cham, Switzerland, 2016. DOI: 10.1007/978-3-319-43162-8. 24

[29] D. Simon, *Evolutionary Optimization Algorithms*, Wiley, Hoboken, NJ, 2013. 26

[30] D. P. Pulsifer, A. Lakhtakia, M. S. Narkhede, M. J. Domingue, B. G. Post, J. Kumar, R. J. Martín-Palma, and T. C. Baker, Fabrication of polymeric visual decoys for the male emerald ash borer (*Agrilus planipennis*), *Journal of Bionic Engineering*, 10:129–138, 2013. DOI: 10.1016/s1672-6529(13)60207-3. 29, 30, 31

[31] M. J. Domingue. A. Lakhtakia, D. P. Pulsifer, L. P. Hall, J. V. Badding, J. L. Bischof, R. J. Martín-Palma, Z. Imrei, G. Janik, V. C. Mastro, M. Hazen, and T. C. Baker, Bioreplicated visual features of nanofabricated buprestid beetle decoys evoke stereotypical male mating flights, *Proceedings of U.S. National Academy of Sciences*, 111:14106–14111, 2014. DOI: 10.1073/pnas.1412810111. 29, 31

[32] D. A. Herms and D. G. McCullough, Emerald ash borer invasion of North America: History, biology, ecology, impacts, and management, *Annual Review of Entomology*, 59:13–30, 2013. DOI: 10.1146/annurev-ento-011613-162051. 30

[33] J. P. Lelito, I. Fraser, V. C. Mastro, J. H. Tumlinson, K. Böröczky, and T. C. Baker, Visually mediated "paratrooper copulations" in the mating behavior of *Agrilus planipennis* (Coleoptera: Buprestidae), a highly destructive invasive pest of North American ash trees, *Journal of Insect Behavior*, 20:537–552, 2007. DOI: 10.1007/s10905-007-9097-9. 30

[34] N. Dushkina and A. Lakhtakia, Structural colors, *Engineered Biomimicry*, A. Lakhtakia and R. J. Martín-Palma, Eds., pages 267–303, Elsevier, Waltham, MA, 2013. DOI: 10.1016/c2011-0-06814-x. 31

[35] M. J. Domingue, D. P. Pulsifer, A. Lakhtakia, J. Berkebile, K. C. Steiner, J. P. Lelito, L. P. Hall, and T. C. Baker, Detecting emerald ash borers (*Agrilus planipennis*) using branch traps baited with 3D-printed beetle decoys, *Journal of Pest Science*, 88:267–279, 2015. DOI: 10.1007/s10340-014-0598-y. 30

[36] T. Gupta, S. E. Swiontek, and A. Lakhtakia, Simpler mass production of polymeric visual decoys for the male emerald ash borer (*Agrilus planipennis*), *Journal of Bionic Engineering*, 12:263–269, 2015. DOI: 10.1016/s1672-6529(14)60118-9. 31

[37] A. D. Briscoe and L. Chittka, The evolution of color vision in insects, *Annual Review of Entomology*, 46:471–510, 2001. DOI: 10.1146/annurev.ento.46.1.471. 32

[38] W. Whewell, On the connexion of the physical sciences, *Quarterly Review*, 51:54–68, 1884. 32

[39] S. Ross, Scientist: The story of a word, *Annals of Science*, 18:65–85, 1962. DOI: 10.1080/00033796200202722. 32

[40] T. Pearce, Convergence and parallelism in evolution: A neo-Gouldian account, *British Journal for the Philosophy of Science*, 63:429–448, 2012. DOI: 10.1093/bjps/axr046. 33

[41] E. Graeff, N. Maranzana, and A. Aoussat, Engineers' and biologists' roles during biomimetic design processes, towards a methodological symbiosis, *Proceedings of the 22nd International Conference on Engineering Design (ICED19)*, pages 319–328, Delft, The Netherlands, Aug. 5–8, 2019. DOI: 10.1017/dsi.2019.35. 33

CHAPTER 4

Rationale for Biologically Inspired Design

Wilful waste makes woeful want.

JAMES KELLY, *A Complete Collection of Scottish Proverbs* (1721)

Since the laws of physics hold sway over every biological process just as completely as over every technological operation, the bioworld should be considered as a repository of answers to billions of technological questions [1]. Some of these answers have already been implemented by humans. Some answers may not be optimal for our technological requirements but can still illuminate possible research directions.

The fact is that the bioworld offers a palette of solutions that may be otherwise unavailable to humans. An example is furnished by three-dimensional photonic crystals with diamond crystal structure, which reflect incident electromagnetic waves in a specific spectral regime, regardless of the direction of incidence [2]. These photonic crystals have been made for operation in the microwave and infrared spectral regimes, but no technique has been successful to fabricate them for operation in the visible spectral regime [3]. Yet the exocuticle of the Brazilian weevil *Lamprocyphus augustus* displays the desired response characteristics in the yellow-green portion of the visible spectral regime [4], as shown in Fig. 4.1. Clearly, then a fabrication route exists in the bioworld that is not known yet to humans.

A review of relevant characteristics of bioworld solutions is undertaken in this chapter to offer the rationale for biologically inspired design.

4.1 ENERGY EFFICIENCY

Neither biological processes nor industrial processes can overcome the fundamental limitations encoded in the laws of physics. Nevertheless, the contrast between the two types of processes is remarkable. Chemical routes are commonplace for material transformations in biological processes, whereas physical routes are routinely employed in industrial processes. This difference is quite succinctly captured by just one environmental variable: temperature.

Consider the temperature differences between biological and industrial processes. Although a few animals live in extreme conditions, the range of temperature in the majority of the bioworld is quite restricted. Deep-sea creatures at 1000 m below sea level have to live at about 5°C, polar bears have to contend with −70°C and desert foxes with 50°C. But the tem-

Figure 4.1: Exocuticle fragment from the Brazilian weevil *Lamprocyphus augustus*.

peratures of internal tissue vary in a much smaller range, because biological cells are mostly water. Accordingly, numerous biological processes occur between, say, 5°C and 45°C. In contrast, very high temperatures are routinely employed in industrial processes. Wood combusts at about 300°C, clay bakes at about 760°C, and iron melts at higher than 1500°C.

The metal zirconium is produced on reducing zirconium chloride by liquid magnesium at about 825°C [5]. The hardness of zirconium on the Mohs scale is 5, the scale ranging from 1 (talc) to 10 (diamond). Tooth enamel, which has the same hardness as zirconium, is formed at a much lower temperature of about 37°C.

The production of high temperatures requires considerable expenditure of energy, implying that biological processes are energy efficient in comparison to industrial processes [6]. This ENERGY EFFICIENCY is a persuasive argument for mimicking biological processes when designing an industrial production line, especially during the time of climate emergency we are presently living in [7]. Indeed, one can justifiably argue that an embrace of biologically inspired design is essential to the survival of the human species as well as numerous other species, in the 21st century on Earth.

4.2 CIRCULAR ECONOMY OF MATERIALS

About 40,000 metric tons of cosmic dust fall on our planet every year [8], but about 50,000 metric tons of hydrogen and helium escape every year too [9]. These changes to the mass of Earth are so tiny that it can be regarded as a closed system wherein materials cycle between the lithosphere, atmosphere, and hydrosphere.

The biosphere comprises parts of each of these three regions of Earth. Biomass, i.e., the mass of living organisms, varies greatly with time [10]. For example, it reduces significantly during the autumn season in the northern hemisphere. Despite such variations, the main outputs, byproducts, and wastes of every living organism become inputs to living organisms of other species. Herbivores eat plants, carnivores eat herbivores, numerous organisms sustain themselves on the excretions and secretions of other species, and the bodies of dead organisms return nutrients to the ground in which plants grow. Leaving aside the sequestration of materials through geological processes, materials thus circulate in the bioworld.

In other words, the bioworld exhibits CIRCULAR ECONOMY [11] of materials, especially when annually averaged over ecologically distinct parts of the biosphere. This circular economy becomes easily evident when an island subspecies are compared to its continental counterpart in average size. Adjustment to serious restrictions on the availability of edible matter on islands is commonly shown by increases and decreases of average sizes of diverse species in relation to their continental cousins [12].

The circular economy of materials evinced by various swathes of the bioworld is not accompanied by the circular economy of energy. This is because our planet is a closed but not an isolated system thermodynamically. A closed system can exchange energy but not mass with its surroundings. An isolated system can exchange neither energy nor mass with its surroundings. The sun supplies energy to Earth, which is in addition to the energy made available to the biosphere by the planetary core.

In the bioworld, every organ is functional over a certain period of time that is, on average, not less than the time needed to reproduce at least once. Many organs are repairable and some organs are not totally necessary for the survival of the individual. The byproducts and waste products of bioworld processes are used as inputs to other bioworld processes, not necessarily in the same organism. Materials in a dead organism provide sustenance to other organisms, either directly or indirectly. Biologically inspired design can influence the manufacture, use, and disposal of specific products with minimal depletion of materials and with minimal impact on the rest of the biosphere; furthermore, energy could be harvested from whatever remains that cannot be cannibalized after use.

4.3   MULTIFUNCTIONALITY

MULTIFUNCTIONALITY is commonplace in living organisms [13–15]. Thus, limbs are used for moving, signaling, gathering and preparing food, wielding weapons, and initiating as well as warding off physical assaults, among other things. Mouths are used for ingesting food and fluids, releasing sounds, breathing, and kissing. As certain organs can perform two or more distinct functions that are not highly related to each other, fewer organs need to be formed and housed in the organism and fewer structures need to be coordinated by the organism's brain.

This economy of multifunctionality is an attractive feature of biologically inspired design [16, 17]. A multifunctional module can be incorporated in a variety of products, thereby reducing inventory costs, enhancing repairability and product lifetimes, and promoting standardization. A multifunctional product may designed and fabricated as an assembly of monofunctional components. A simple example is a Swiss Army knife. A multifunctional product could also be made from multifunctional materials, whether natural or composite. The costs of eventual disposal may be higher when composite materials are used, and designers will have to make choices based on lifecycle audits [18].

4.4   MULTICONTROLLABILITY

The concept of MULTICONTROLLABILITY [19] is closely allied to multifunctionality. Multicontrollability is also exhibited commonly in the bioworld. Thus, multiple modes of locomotion can be used by an organism to propel itself from one location to another, and often the same sound can be uttered using two or three different placements of the tongue in the buccal cavity. We get alarmed by hearing the sound of an approaching car and/or by seeing it. Reliance on multiple mechanisms thus builds resilience via redundancy. That's why multiple control modalities are used to ensure specific actions in critical facilities such as nuclear power plants and missile guidance centers.

4.5   SUBOPTIMALITY

When mimicking a bioworld product or process, it is important to remember that biological phenomena are adapted to a specific context with a given set of constraints. This means that the solutions derived from a biological phenomenon may not be suitable in contexts with different constraints. For instance, the wings of an owl are silent but are unsuitable for rapid flight, the wings of a swan are noisy but can lift a heavy body, and the wings of a swift allow for very high speed but make it very difficult for the bird to take off from the ground.

A bioworld solution is also constrained by evolutionary history since it arises from successive mutations of several species [20]. Each mutation could be suboptimal that performs just well enough in a particular niche. A succession of such mutations will definitely produce a solution that too is viable in its niche, but that solution could be suboptimal even in that niche.

SUBOPTIMALITY in the bioworld has long been exemplified by the plethora of visual problems that plague humans [21], not to mention other mammals. Aberrations, astigmatism, and blindspots are structural deficiencies that have kept generations of ophthalmologists gainfully employed. Although all of their patients would like to keep using their eyes for as long as possible, the human eye can hardly be regarded as the product of a well-designed instrument [22].

As a bioworld solution is not necessarily optimal even in the bioworld, it is likely to require some modification to optimize it for a specific technoscientific application. This should be viewed as a welcome opportunity, all the more so as the need for modification may allow the incorporation of functionalities not associated with the bioworld solution in the bioworld. Thus, the rapidity of action of biomimetic insulin can be controlled by the alteration of the codon sequence, as mentioned in Section 3.3.2. Similarly, bioreplicated decoys can be colored differently from the the species being replicated, as discussed in Section 3.3.3.

Further opportunities may arise after realizing that several bioworld solutions can be combined for a specific technoscientific application. This is exemplified by the tennis racquets in the Dunlop Biomimetic 200™ series. The racquet beam is made of Dunlop HM6 carbon sandwiched between aerogel-enhanced carbon sheets. Dunlop HM6 carbon mimics the morphology of honeycombs which are extraordinarily strong and lightweight structures [23]. The surface of the racquet frame is covered by a fabric with overlapping scale-like protrusions to reduce aerodynamic drag. These protrusions mimic denticles that reduce hydrodynamic drag and prevent fouling of shark skins [24, 25]. The surface of the racquet grip mimics the setae on the feet of a gecko that enable it to walk upside down on smooth surfaces [26, 27].

4.6   CONTRAINDICATED PERFORMANCE

For over two millennia, humans have known that an object denser than water sinks in a bathtub but an object of lesser density than water floats. Well, boats float in rivers and seas, but that is because the volume-averaged density of a boat's hull and superstructures as well as of air below the waterline is the same as of water.

Air is a liquid and a rigorous scientific study [28] is not needed to prove that a bird is definitely heavier than air on a unit-volume basis. Although avian flight is thus contraindicated, birds of most species can fly well, some even at altitudes higher than 10 km [29]. The secret lies in the arrangement of flight feathers arranged on concave wings that can be flapped to raise the underwing pressure and provide lift.

Mushrooms and their mycelium roots are well known to be very fragile. But a fungus growing in a fibrous material functions as a glue that provides the resulting composite material with surprisingly high stiffness and strength. This can be seen in the forest floor where the soil in places with fungus can become harder and stiffer, provided the soil is a good mixture of organic material of various sizes. The same phenomenon can be utilized for making building components and plates from straw by letting a fungus grow in the humidified material. The mycelium roots will bind the straw fibers together and form a stiff composite material, as depicted in Fig. 4.2. Both foams and structural composites are being made of mushrooms [30, 31].

Mollusk shells are calcareous, created by the secretion of calcium carbonate mixed in a broth of polysaccharides and glycoproteins which controls the position and elongation of calcium-carbonate crystals [32]. As talc, calcium carbonate is among the softest natural materials known. As aragonite, the material's hardness does not exceed 4 on the Mohs scale. Yet, mollusk shells comprising interlaced plates of aragonite are extremely durable, with a modulus of elasticity similar to wood's, tensile strength similar to copper's, and compressive strength higher than porcelain's [33]. The secret lies in the arrangement of aragonite plates that prevents crack propagation and thereby provides the toughness needed to protect the enclosed body. The same arrangement of plates of Norwegian slate has been used in the retaining walls constructed on the undulating terrain of the Lyngby campus of Danmarks Tekniske Universitet (DTU) as shown in Fig. 4.3.

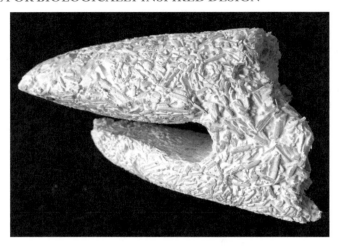

Figure 4.2: Mycelium bio-composite made from straw and other agricultural byproducts.

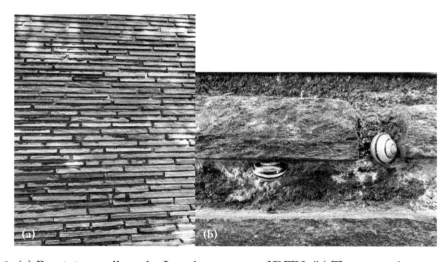

Figure 4.3: (a) Retaining wall on the Lyngby campus of DTU. (b) The inter-plate regions of the wall provide habitat for terrestrial mollusks of the species *Cepaea nemoralis*.

The examples of CONTRAINDICATED PERFORMANCE in the bioworld offer unexpected routes to the seemingly impossible satisfaction of mutually incompatible constraints. Thus, biologically inspired design has the potential to engender innovative products and processes.

4.7   REFERENCES

[1] V. Davidov, Biomimicry as a meta-resource and megaproject, a literature review, *Environment and Society: Advances in Research*, 10:29–47, 2019. DOI: 10.3167/ares.2019.100103. 37

[2] M. Maldovan and E. L. Thomas, Diamond-structured photonic crystals, *Nature Materials*, 3:593–600, 2004. DOI: 10.1038/nmat1201. 37

[3] A. Risbud, A. Lakhtakia, and M. H. Bartl, Towards bioreplicated texturing of solar-cell surfaces, *Encyclopedia of Nanotechnology, Part 20*, pages 2755–2762, B. Bhushan, Ed., Springer, Heidelberg, Germany, 2012. DOI: 10.1007/978-90-481-9751-4_18. 37

[4] J. W. Galusha, L. R. Richey, J. S. Gardner, J. N. Cha, and M. H. Bartl, Discovery of a diamond-based photonic crystal structure in beetle scales, *Physical Review E*, 77:050904, 2008. DOI: 10.1103/physreve.77.050904. 37

[5] L. Xu, Y. Xiao, A. van Sandwijk, Q. Xu, and Y. Yang, Production of nuclear grade zirconium: A review, *Journal of Nuclear Materials*, 466:21–28, 2015. DOI: 10.1016/j.jnucmat.2015.07.010. 38

[6] J. F. V. Vincent, O. A. Bogatyreva, N. R. Bogatyrev, A. Bowyer, and A.-K. Pahl, Biomimetics: Its practice and theory, *Journal of the Royal Society Interface*, 3:471–482, 2006. DOI: 10.1098/rsif.2006.0127. 38

[7] P. Gilding, Why I welcome a climate emergency, *Nature*, 573:311, 2019. DOI: 10.1038/d41586-019-02735-w. 38

[8] H. A. Zook, Spacecraft measurements of the cosmic dust flux, *Accretion of Extraterrestrial Material Throughout Earth's History*, B. Peucker-Ehrenbrink and B. Schmitz, Eds., pages 75–92, Springer, New York, 2001. DOI: 10.1007/978-1-4419-8694-8_5. 38

[9] D. C. Catling and K. J. Zahnle, The planetary air leak, *Scientific American*, 300(5):36–43, 2009. DOI: 10.1038/scientificamerican0509-36. 38

[10] R. A. Houghton, Biomass, *Encyclopedia of Ecology*, S. E. Jørgensen and B. D. Fath, Eds., pages 448–453, Elsevier, New York, 2008. 38

[11] W. R. Stahel, *Circular Economy: A User's Guide*, Routledge, Abingdon, Oxford, UK, 2019. 39

[12] J. B. Foster, Evolution of mammals on islands, *Nature*, 202:234–235, 1964. DOI: 10.1038/202234a0. 39

[13] D. H. Evans, P. M. Piermarini, and K. P. Choe, The multifunctional fish gill: Dominant site of gas exchange, osmoregulation, acid-base regulation, and excretion of nitrogenous waste, *Physiological Reviews*, 85:97–177, 2005. DOI: 10.1152/physrev.00050.2003. 39

[14] S. N. Patek, J. E. Baio, B. L. Fisher, and A. V. Suraez, Multifunctionality and mechanical origins: Ballistic jaw propulsion in trap-jaw ants, *Proceedings of U.S. National Academy of Sciences*, 103:12787–12792, 2006. DOI: 10.1073/pnas.0604290103. 39

[15] D. M. Neustadter, R. L. Herman, R. F. Drushel, D. W. Chestek, and H. J. Chiel, The kinematics of multifunctionality: Comparisons of biting and swallowing in *Aplysia californica*, *Journal of Experimental Biology*, 210:238–260, 2007. DOI: 10.1242/jeb.02654. 39

[16] A. Lakhtakia, From bioinspired multifunctionality to mimumes, *Bioinspired, Biomimetic and Nanobiomaterials*, 4:168–173, 2015. DOI: 10.1680/jbibn.14.00034. 39

[17] A. Lakhtakia and W. Johari, Engineered multifunctionality and environmental sustainability, *Journal of Environmental Studies and Sciences*, 5:732–734, 2015. DOI: 10.1007/s13412-015-0305-1. 39

[18] D. F. Ciambrone, *Environmental Life Cycle Analysis*, CRC Press, Boca Raton, FL, 1997. DOI: 10.1201/9780203757031. 40

[19] A. Lakhtakia, D. E. Wolfe, M. W. Horn, J. Mazurowski, A. Burger, and P. P. Banerjee, Bioinspired multicontrollable metasurfaces and metamaterials for terahertz applications, *Proceedings of SPIE*, 10162:101620V, 2017. DOI: 10.1117/12.2258683. 40

[20] D. Adriaens, Evomimetics: The biomimetic design thinking 2.0, *Proceedings of SPIE*, 10965:1096509, 2019. DOI: 10.1117/12.2514049. 40

[21] H. Helmholtz, *Popular Lectures on Scientific Subjects*, Appleton, New York, 1885. DOI: 10.1037/12825-000. 40

[22] R. S. Fishman, Darwin and Helmholtz on imperfections of the eye, *Archive of Ophthalmology*, 128:1209–1211, 2010. DOI: 10.1001/archophthalmol.2010.189. 40

[23] T. Blitzer, *Honeycomb Technology: Materials, Design, Manufacturing, Applications and Testing*, Chapman and Hall, London, UK, 1997. DOI: 10.1007/978-94-011-5856-5. 41

[24] G. D. Bixler and B. Bhushan, Biofouling: Lessons from nature, *Philosophical Transactions of the Royal Society of London A*, 370:2381–2417, 2012. DOI: 10.1098/rsta.2011.0502. 41

[25] T. Sullivan and F. Regan, The characterization, replication and testing of dermal denticles of *Scyliorhinus canicula* for physical mechanisms of biofouling prevention, *Bioinspiration and Biomimetics*, 6:046001, 2011. DOI: 10.1088/1748-3182/6/4/046001. 41

[26] K. Autumn and A. M. Peattie, Mechanisms of adhesion in geckos, *Integrative and Comparative Biology*, 42:1081–1090, 2002. DOI: 10.1093/icb/42.6.1081. 41

[27] C. Majidi, R. E. Groff, Y. Maeno, B. Schubert, S. Baek, B. Bush, R. Maboudian, N. Gravish, M. Wilkinson, K. Autumn, and R. S. Fearing, High friction from a stiff polymer using microfiber arrays, *Physical Review Letters*, 97:076103, 2006. DOI: 10.1103/physrevlett.97.076103. 41

[28] T. W. Seamans, D. W. Harnershock, and G. E. Bernhardt, Determination of body density for twelve bird species, *Ibis*, 137:424–428, 1995. DOI: 10.1111/j.1474-919x.1995.tb08046.x. 41

[29] R. C. Laybourne, Collision between a vulture and an aircraft at an altitude of 37,000 feet, *The Wilson Bulletin*, 86:461–462, 1974. https://www.jstor.org/stable/4160546 41

[30] E. Bayer and G. McIntyre, *Method for making dehydrated mycelium elements and product made thereby*, US Patent 2012/0270302 A1, October 25, 1997. https://patents.google.com/patent/US20120270302A1/en 41

[31] C. Bruscato, E. Malvessi, R. N. Brandalise, and M. Camassola, High performance of macrofungi in the production of mycelium-based biofoams using sawdust—sustainable technology for waste reduction, *Journal of Cleaner Production*, 234:225–232, 2019. DOI: 10.1016/j.jclepro.2019.06.150. 41

[32] F. Marin and G. Luquet, Molluscan biomineralization: The proteinaceous shell constituents, of *Pinna nobilis* L., *Materials Science and Engineering C*, 25:105–111, 2005. DOI: 10.1016/j.msec.2005.01.003. 41

[33] F. Barthelat, Nacre from mollusk shells: A model for high-performance structural materials, *Bioinspiration and Biomimetics*, 5:035001, 2010. DOI: 10.1088/1748-3182/5/3/035001. 41

CHAPTER 5

Problem-Driven Biologically Inspired Design

It is, that as existing human inventions have been anticipated
by Nature, so it will surely be found that in Nature lie the prototypes of inventions not yet revealed to man. The great discoverers of
the future will, therefore, be those who will look to Nature for Art,
Science, or Mechanics, instead of taking pride in some new invention,
and then finding that it has existed in Nature for countless centuries.
REV. JOHN G. WOOD, *Nature's Teachings, Human Invention
Anticipated by Nature* (1877)

5.1   INTRODUCTION

Biologically inspired design (BID) can be approached from two different directions [1–3]. The approach from the engineering side is referred to as PROBLEM-DRIVEN BID, whereas the approach from the biology side leads to SOLUTION-DRIVEN BID. The former is treated in this chapter, the latter in Chapter 6.

As the name implies, problem-driven BID is initiated by an engineering problem whose solutions are sought; hence, it is very similar to traditional engineering design. The major difference is that the solution principles are searched in the bioworld. As engineering designers will be familiar with the design-oriented parts of the process but are likely to be less knowledgable and experienced in the tasks that relate to biology, problem-driven BID should be carried out in a collaboration between engineers and biologists. However, there are strong limitations for problem-driven BID in such a collaboration, as explained in Sections 5.2.2–5.2.4.

PROBLEM-DRIVEN BID is the term used by researchers at the Georgia Institute of Technology [1], Arts et Métiers ParisTech [4, 5], and Danmarks Tekniske Universitet (DTU) [2]. The International Standards Organization calls it TECHNOLOGY-PULL BIOMIMETICS because a technological need initiates it and drives the work [6]. The term TOP-DOWN BIONIK has been used by researchers at the Technische Universität München for many years [7]. It is also this type of BID that is handled with the DESIGN SPIRAL from the Biomimicry Institute [8].

There are other ways than problem-driven BID to generate new ideas for how to design products and other artifacts. One can look at already existing products or even search patents. Or

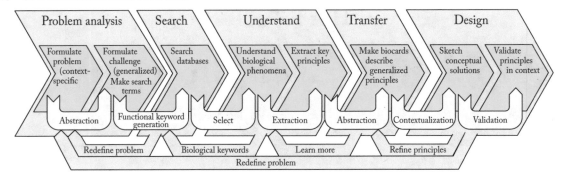

Figure 5.1: The five phases of problem-driven BID implemented using the DTU biocard method.

one could turn to a range of different creativity techniques such as brainstorming, 635-method or the Scamper method [9, 10]. Two questions naturally arise. First, how well does BID perform as an idea-generation technique? Second, are its outcomes worth the effort?

Answers to these questions have been sought by comparing the BID methodology to traditional brainstorming [11]. Several design students were given an assignment to generate ideas to a given problem, with half of the students asked to use brainstorming and the other half to use the BID methodology. The novelty of each resulting design proposal was identified by comparing it with other solutions found on the internet. The comparison was made using the SAPPhIRE model for causality [12] where the similarity between new and existing design proposals was compared at seven levels of abstraction. The use of BID methodology resulted in fewer design proposals, but the ones that were found were more novel (and, therefore, presumably of higher quality). This is a key argument for using the BID methodology. Brainstorming is easy to learn and requires little preparation or skills, thereby producing many design proposals. On the contrary, the BID methodology requires a stricter procedure to be followed as well as some interest in and some knowledge of biology, but results in novel proposals.

5.2   PHASES OF PROBLEM-DRIVEN BID

Implementation of problem-driven BID is done in five phases, beginning with an initial analysis of the design problem, followed by a search for biological analogies, then distilling an understanding of biological phenomena to extract key principles, followed by a reformulation of design principles, and ending with the actual design of new objects after validating the principles in the context of the design problem. The flow chart in Fig. 5.1 illustrates these five phases using the biocard method developed at DTU.

Abstraction is done at least twice during the problem-driven BID process, as is clear from Fig. 5.2. First, the technical problem is abstracted from the initial analysis of the design

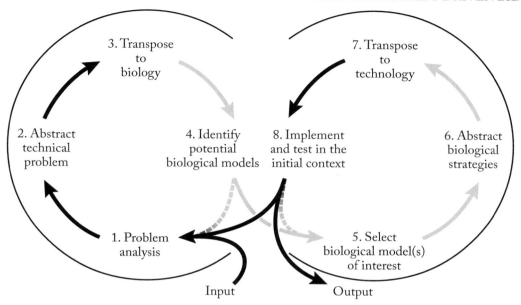

Figure 5.2: Interaction of technology and biology in problem-driven BID [5].

problem. Second, key biological principles or strategies are abstracted from the understanding of a biological phenomenon and brought into a form useful for design work.

5.2.1   FIRST PHASE: PROBLEM ANALYSIS

The first phase in problem-driven BID is no different from those in numerous other goal-oriented projects, since a thorough understanding of the problem being tackled is required. Asking the right question(s) is halfway to finding good solutions. The problem-analysis phase can involve just a single person but is often better carried out as a collaboration of several people. Discussions among team members force clarity in the description of the problem so that every member can get a clear and complete picture. The following tasks have to be undertaken in the problem-analysis phase: problem description, function analysis, and engineering-biology translation.

Problem-Description Task
Describing the design problem adequately is among the most important activities in BID, just as it is in design work in general. Adequate understanding of the core issues and the short-comings of existing products determines the form and the substance of the remainder of the design process. Understanding the design problem and describing it clearly for others can be difficult enough for a single person, but it becomes even more complicated when many persons

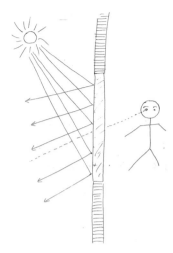

Figure 5.3: A hand-drawn sketch describing the window problem. The window must allow an external view but prevent the solar infrared radiation from entering the room.

collaborate in a design team. Therefore, the problem must be described and communicated in a way that it is easily and uniformly understood by many persons. A sequence of illustrations, whether drawn by hand or on computers, accompanied by bulleted points in text can document the problem reasonably well. Illustrations can be rapidly made and transcend barriers of language, terminology, and expertise. The technical problem can then be abstracted quite easily.

But, care must be taken that the illustrations focus on the desired functionality but not on the manner in which the problem is to be solved. As an example, consider the window problem that architects often encounter when designing buildings in normally sunny locales. People inside a building are interested both in having sunlight enter rooms through windows and in being able to view the outside. However, solar radiation contains not only visible light but also infrared waves that heat the room and may necessitate the increased use of air-conditioning systems. The design problem is that of a window that allows the occupants of a room to enjoy the external view but (partially) prevents solar infrared radiation from entering the room. This design problem can be described by the simple sketch shown in Fig. 5.3. The window pane is represented by two parallel lines, the external view is illustrated by a dashed straight line that begins at one eye of a stickperson and crosses the window pane to the exterior, and the infrared restriction is represented by bouncing arrows. Such an abstract description will stimulate an open-minded approach to identify the core functionality and allow for a broader and goal-oriented search for biological organisms displaying that functionality.

Another method that is useful for problem analysis and description is the four-box method [13]. This method requires the design team to specify

Operational environment	Functions
Specifications	Performance criteria

Figure 5.4: The four-box method for problem analysis and description [13].

   (i) the operational environment for the product (i.e., the context),

  (ii) the core functions delivered by the product,

 (iii) the main specifications of the product, and

  (iv) the performance criteria that the product must satisfy.

The responses are entered as bulleted lines of text in the table shown in Fig. 5.4. For the window example, the operational environment includes the type of room in which the window is to placed (i.e., office, school room, bed room, etc.) as well as the geographical location and climatic conditions (e.g., dry/humid, sandy/salty, hot/cold, etc.). Functions could include "provide transparency," "prevent solar infrared radiation to pass through," and "allow cleaning;" see also Section 5.2.1. Specifications include linear and areal dimensions and orientation toward the sun in summer. Performance criteria could include the fraction of visible light that is allowed to pass through the window, the color tint that is acceptable, and the minimum acceptable viewing angle.

Function-Analysis Task
A problem is typically specified using a terminology which is closely related to the context of the problem. For instance, will a car driver explain a puncture in a tire as "having a flat tire"? However, as described in Section 2.4.1, it is important to provide an abstract functional description rather than a concrete one, in order to prevent fixation. The puncture problem can, of course, be solved by changing the tire; but if the goal is to prevent punctures, it is advantageous to describe the function in more abstract terms. The tire is a solution to the functions "provide road grip" and

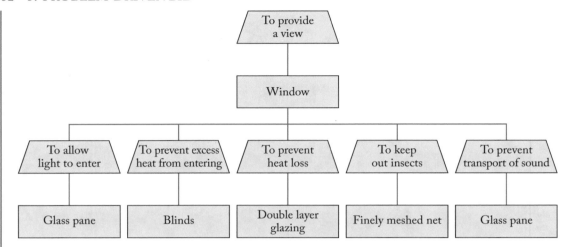

Figure 5.5: Functions-means tree diagram for a window. Each trapezoidal block contains a function, each rectangular block a means.

"provide driving comfort." By broadening the problem description using such abstract terms, it is more likely that a completely different solution will be found. The road grip could be provided by spiked solid wheels and the comfort could be supplied by a sturdy mechanism for wheel suspension. Such a wheel solution will not suffer from punctures.

More generally, an engineering problem can be analyzed by describing an artifact that solves the problem. The artifact can be decomposed into functional units each of which is described in terms appropriate for the context. The next step is then to formulate the function(s) of each artifact with a more abstract terminology that allows for a broader search for alternative means to solve the problem. The overall problem is decomposed into sub-functions, each describing specific aspects of what the artifact does and defining a set of metrics for the required performance.

Function analysis for the window problem of Fig. 5.3 can be performed as follows. The main function of a window is to provide a view. This can be done with glass panes, but an open hole in the wall will also deliver this function. A functions-means tree diagram, as described in Section 2.4.2, helps to define which functionalities are required and thus support a search for alternative solutions. Figure 5.5 shows a functions-means tree diagram for the window problem with each trapezoidal box containing a function or sub-function and each square box containing a means to provide the needed functionality. The search for solutions is thus broken down into identification of various means, each of which solves a specific aspect of the overall problem. The top-level functionality in the functions-means tree diagram is "to provide a view." The main function can be broken down into five sub-functions: "to allow light to enter," "to prevent excess heat from entering," "to prevent heat loss," "to keep out insects," and "to prevent sound transport." The sub-function involving insects can be solved by using a finely meshed net as an alternative to a glass pane. The last sub-function rules out a hole as a window and also the finely meshed net. The functions-means tree diagram therefore is a tool for qualifying the search for solutions and it is also very helpful in the search for analogous solutions from the bioworld.

A challenge in describing functionalities for the functions-means tree diagram is to select the right phrases that will be helpful in the search phase. Assistance can be taken from on-line thesauri wherein synonyms and antonyms can be found [14]. Another helpful resource is the WordNet database from Princeton University [15].

Engineering-Biology Translation Task

After a designer (or a design team) has described the problem and identified the desired functionalities, the formulation is often very technical. This is a good starting point, since the designer should be familiar with the engineering terminology and therefore should be able to formulate the problem precisely enough to find good technical terms for searching the literature.

In Fig. 5.1, this task is referred to as the context-specific formulation of the design problem. However, it is not very likely that exactly the same terms are used to describe similar functions in the engineering and biology literatures. Before searching in the biology literature, it is therefore beneficial to translate the engineering terms to biology terms. This task can be approached by looking at synonyms in a thesaurus as well as by looking in those segments of the biology literature wherein similar phenomena are likely to be found. If, for instance, a new type of cleaning mechanism is to be designed, then one could consult the literature on how domesticated animals as well as animals housed in diverse research institutions (such as zoological parks) keep themselves tidy. In that literature, terms such as "washing," "licking," and "removing hair and dirt" are used instead of "cleaning." The terms from biology literature could be more useful in finding similar phenomena among other animals. The abstraction activity of finding good biological search terms is referred to as the formulation of generalized challenges in Fig. 5.1.

For the window example, technical search terms could be "semi-transparent," "sun blocking," and "shield light." These would only find a few biological analogies but that is a good starting point. Once the first biological analogy has been identified, the biology literature could be consulted to find out what terms are used to describe protection from high-intensity light. As animal eyes are likely to possess features for such protection, literature on veterinary ophthalmology would be appropriate. Animals protect their eyes from high-intensity light by contracting the iris, closing eye lids, moving the eyelashes, and using skin folds that shade. Another search could be for plants growing in sunny deserts, because those plants somehow avoid being overheated, e.g., how cacti utilize corrugated surfaces and spines for cooling by convection. The insight gained could then be used to define search terms more likely to be found in the biology literature. Examples of search terms could be "eye protection" and "temperature regulation."

Another helpful approach is to translate the terms used for biological phenomena into Latin. Latin words are universally used in the scientific literature, especially in the biology literature. Taxonomists use Latin terms for kingdoms, phyla, classes, orders, suborders, families, genera, and species, each term usually referring to a specific biological attribute. After a Latin term is found in taxonomy, it is straightforward to move up, down, or sideways in the hierarchy to find other organisms and then explore other bioworld solutions to the design problem.

Researchers at the University of Toronto have developed a natural search approach for BID and a method for identifying good biological search terms. They have proposed a set of techniques for abstraction and identification of relevant search terms to be used for the biological search [3]. Verbs are recommended instead of nouns since it is more likely that nouns will lead to pre-conceived analogies. Furthermore, verbs describe actions and hence are better for finding a greater variety of biological forms. As an example, the verb "protect" helps find a greater variety of phenomena than the noun "cuticle" does. If certain verbs that can be considered as biologically meaningful (significant or connotative) occur more commonly than others, they can be considered as bridge words that are more likely to be helpful in the biological search.

5.2.2 SECOND PHASE: SEARCH

Searches can be done in many different ways. Most straightforwardly today, internet search engines such as Yahoo, Google, and Bing should be used. The challenge is that, as no search engine is restricted to biology, a large number of hits will result that can be difficult to navigate through. It is therefore important to identify a good starting point when using an internet engine. One way is to apply a bio-brainstorm where the person or groups of persons formulate a question of following kind: "How would this particular problem be overcome in the bioworld?" Based on the biological knowledge already available in the design team, animals and plants can be identified. For instance, many people would readily propose mammalian eyes as biological solutions to the window problem. This first hit will be a good starting point for a wider search.

Another approach is to use dedicated biology databases that will be more likely to propose relevant biological organisms. Among the better known databases is *AskNature* developed by the Biomimicry Institute [16]. AskNature contains a large number of examples of biomimicry, with biological organisms described alongside how a biological strategy has been transferred into technical applications. AskNature provides at least two ways to initiate a search. One is a simple free-text search very similar to the use of internet search engines. Another is to use a biomimicry taxonomy [17] which describes functions on three hierarchical levels: group, subgroup, and function. Relevant for the window example could be to focus on the group "protect from physical harm," the subgroup "protect from non-living threats," or the function "(protect from) light."

The terms from the biomimicry taxonomy can be used not only for a focused search in AskNature but also when searching more broadly in other databases or on the internet. As additional search terms are needed to limit an internet search to biological phenomena, terms such as "biology," "animal," and "plant" or other biology-related terms should be added. The right terms must be found through an iterative approach where relevant hits can be used to identify relevant biological terms that will guide the search in a fruitful direction.

Yet another approach is to use library search engines to search scientific books and papers. The library databases are prepared to offer goal-directed searchs where the focus is on recognized and quality-tested scientific knowledge. What is found using library search engines therefore has a high degree of trustworthiness. The difficulty in using scientific literature is that is written in the language of a sub-culture, i.e., it can be difficult for a layperson to understand a paper written for a specialist journal or book.

There are also other ways to search for biological organisms with relevant mechanisms and functionalities. An obvious one is to consult a professional biologist. They have broad insights about the bioworld, they know how many biological organisms function, and they can easily peruse scientific literature to gather further information. However, as they may require payment for their services, the value of conducting a biological search must be higher than the cost of hiring a biologist. Besides, a limitation is the growing specialization within the broad discipline of biology. Many biologists today have deep knowledge of only a narrow sub-discipline and therefore are less suited for the broad search for biological phenomena that could help solve a specific design problem.

Finally, there is the option to visit some parts of the bioworld. Once a mind is tuned to looking for a functionality, it is natural to wonder about the things that we see in a forrest, a zoological park, a botanical garden, or a protected area set aside as a nature reserve [18]. For instance, if one is searching for new strategies for bearing structural loads (e.g., columns for holding motorway signs, large tents, or bridges), it is natural to wonder about how trees are structured and anchored in the ground so they can resist high wind pressure in storms. Or, if one is look for self-cleaning strategies, one will find many plants that stay clean despite dirty surroundings.

A possible pitfall when searching for biological phenomena is that only well-known ones are explored. Experiences from teaching BID courses show that many students limit their searches to the larger animals, i.e., mammals, birds, and insects [19]. By limiting the search to the more familiar fauna and flora, the probability of finding really novel ideas decreases. If the search is forced to be broader to cover items such as marine life, microbiology, and single-cell organisms, more and novel ideas emerge [19].

5.2.3   THIRD PHASE: UNDERSTAND

Once a list of promising biological phenomena has been created, the next step is to understand the underlying mechanisms. The mechanism is straightforward to understand in some cases, but not for all. See, for example, Section 3.3.2 for the complexity of insulin production in the human pancreas. It can also be that the overall functionality is easy to understand but becomes more complex after additional detail is required for implementation. For the window example, it is easy to understand that the iris in a mammalian eye functions like a camera aperture with the size of the hole determining how much light is allowed through, but the activation of muscles causing the contraction and widening of the iris is more complicated for a non-specialist to understand.

Better understanding normally requires access to trustworthy literature which can inform about a particular biological phenomenon and explain the underlying mechanism(s) in adequate but not overwhelming detail. Whereas internet searches will supply the needed insight in some cases, a proper library search is necessary more often than not. Relevant keywords and descriptive names of the biological phenomenon combined with boolean operators (AND, OR, NOT) will help identify relevant books and journal papers that can retrieved though the library facilities. Latin terms will be especially useful in library searches since they precisely define the type of biological phenomenon that is described, thereby offering the opportunity to select a more general level in biological taxonomy and find literature for a wider group of organisms.

It can be advantageous to use a dedicated database such as *BIOSIS Previews* [20] at the library. Another useful tool is the *Encyclopedia of Life* [21], a community-driven resource to which many biologists worldwide supply information about animals and plants. A supplementary valuable resource are the biologists themselves. If approached correctly and politely, they will often help with basic explanations and guide toward the relevant literature for deeper understanding.

5.2.4   FOURTH PHASE: TRANSFER

In the next phase of BID, the findings must be transferred to the design problem by describing the underlying functional principle of each biological phenomenon found. This is important to facilitate precise and accurate communication among the members of the design team. If the findings are communicated too loosely, much is left to interpretation and the final design may be inspired by something other than what was intended by the person(s) who found a relevant biological phenomenon.

One way to document the findings is to use biocards [22], an example of which is presented in Fig. 5.6. The figure shows two similar yet different biocards on the mechanism that keeps equine eyes clean: a concrete description using biological terminology and graphics in the left biocard but an abstract description using neutral non-biological terms and graphics in the right biocard. The biocard on the left mentions a tear film to which dust particles adhere, that is removed periodically removed by the eyelid, and which is replenished periodically by tears. This description is suitable for designers to generate ideas, but its scope is limited compared to the abstract description in the biocard on the right. Terms such "tears" and "liquid" will fixate the designer in thinking only of solutions that rely on a liquid to collect and clean. The abstract description replaces both of those terms by the more neutral "substance." This will make it more likely for the designer to think freely and consider both liquid and solid substances for collecting dirt particles. The same argument applies to the graphics in the biocard. Drawings of bioworld solution should be eschewed in favor of more symbolic drawings.

Horse eyeball cleaning
Organism: *Oculus, Equus ferus*

Biological phenomenon: Horse eyes are protected from dirt by a liquid film covering the eye. Dirt particles adhere to the film. The liquid is regularly removed by the flapping of the eyelid and replenished with fresh supply from the lacrimal gland.

Functional principle: (1) The eye is covered with a tear film to which dust particles can adhere. (2) The film and the dust particles are removed mechanically from the eye, and (3) tears replenish the film.

References: Braun, R.J. and Fitt, D. (2003). To minimize shaer stress and to avoid solid to solid contact between the eyelid and the eye surface, the latter is covered by a thin tear film. *Mathematical Medicine and Biology,* 20, 1-28.
Walls, G.L. (1942) The vertebrate eye and its adaptive radiation. New Yourk, Haftner Publishing Company.

Picture credit: Wikipedea.org, Waugsberg

Horse eyeball cleaning
Organism: *Oculus, Equus ferus*

Biological phenomenon: Horse eyes are protected from dirt by a liquid film covering the eye. Dirt particles adhere to the film. The liquid is regularly removed by the flapping of the eyelid and replenished with fresh supply from the lacrimal gland.

Functional principle: (1) A surface is covered with an intermediary substance to which unwanted matter can adhere. (2) The intermediary substance and the unwanted matter can be removed mechanically from the surface, and (3) the removed substance can be replaced by a fresh supply.

References: Braun, R.J. and Fitt, D. (2003). To minimize shaer stress and to avoid solid to solid contact between the eyelid and the eye surface, the latter is covered by a thin tear film. *Mathematical Medicine and Biology,* 20, 1-28.
Walls, G.L. (1942) The vertebrate eye and its adaptive radiation. New Yourk, Haftner Publishing Company.

Picture credit: Wikipedea.org, Waugsberg

Figure 5.6: (left) Concrete and (right) abstract descriptions in a biocard. The biocard on the right is better suited for problem-driven BID.

5.2.5   FIFTH PHASE: DESIGN

The biocards can be used in different ways in the fifth phase of BID. One way is to make a collection of biocards describing different functional principles based on different biological phenomena. The designer or design team can then take one card at a time and sketch solutions based on the functional principle in that biocard. It is important not to evaluate the quality of that principle but wait until a design proposal utilizing it has emerged. In that way, it will be the physical embodiment in a given context that will be evaluated.

For each promising design proposal, a physical model should be constructed for demonstration in order to convince decision makers about investing resources needed to better investigate the proposal. Students taking a BID course at DTU routinely build such proof-of-principle models [23]. Figure 5.7 shows an example. The design problem is that of reducing drag on a ship and thereby lower energy consumption. The model ship in the figure is inspired by emperor penguins [24]. When threatened, an emperor penguin releases air from underneath its feathers. The resulting air bubbles form a thin layer that encapsulate its body to drastically reduce friction. The penguin then increases its speed several times and escapes its enemies by rocketing out of the water to the ice flakes where it will be safer. To prove the air-bubble principle for drag reduction,

Figure 5.7: Inspired by the use of air bubbles by emperor penguins to reduce friction in water, this toy ship as a physical model demonstrated that the same functional principle will reduce drag on a full-size ship. Courtesy: David Maage, Enzo Hacquin, and Anders Lui Soerensen.

a student team made a toy ship and equipped it with two aquarium pumps. On pumping air in tubes with tiny holes underneath the toy ship, its bottom and sides were surrounded by a layer of air bubbles. Measurements of the drag resistance confirmed that a reduced force was needed to propel the toy ship.

5.3   ENGINEERS AND BIOLOGISTS

Since BID is basically about transferring biological knowledge to the engineering domain, it seems obvious to carry out the design work as a collaboration of people with the two competences. There are good examples of successful and sustained collaborations. For instance, Julian Vincent is a biologist who has worked for many years at an engineering college.

Biologists employed at agricultural universities are oriented toward developing more efficient techniques for agriculture and forestry. Although endowed by their education with deep insights into biology, they are oriented toward development work in institutions where the results are solutions that keep the citizenry well fed. Furthermore, their orientation must be sufficiently broad to encompass both botany and zoology.

In contrast, typical faculty members in a biology department are highly specialized, because they achieve professional rewards by focusing on narrow topics within sub-disciplines such as entomology, mycology, and molecular biology. Although beneficial for conducting novel research on narrow topics, that outlook poses a challenge to engineering-biology collaborations. When an engineering designer searches the bioworld to find applicable biological strategies, those strategies may have to be searched through the length and breadth of available biological knowledge. If the biologist in the collaboration is a marine biologist, they will have little knowledge about strategies involving insects or mountain plants.

A research group in Paris examined the role of biologists in biologically inspired design [25]. They found that mixed teams are more effective in coming up with more ideas and make fewer mistakes. They also found that there is an increase in the diversity of biological strategies identified as potentially useful in design work.

5.4   REFERENCES

[1] M. Helms, S. S. Vattam, and A. K. Goel, Biologically inspired design: Process and products, *Design Studies*, 30:606–622, 2009. DOI: 10.1016/j.destud.2009.04.003. 47

[2] T. A. Lenau, A.-L. Metze, and T. Hesselberg, Paradigms for biologically inspired design, *Proceedings of SPIE*, 10593:1059302, 2018. DOI: 10.1117/12.2296560. 47

[3] L. H. Shu, K. Ueda, I. Chiu, and H. Cheong, Biologically inspired design, *CIRP Annals—Manufacturing Technology*, 60:673–693, 2011. DOI: 10.1016/j.cirp.2011.06.001. 47, 54

[4] P.-E. Fayemi, N. Maranzana, A. Aoussat, and G. Bersano, Bio-inspired design characterisation and its links with problem solving tools, *Proceedings of DESIGN: 13th International Design Conference*, pages 173–182, Dubrovinik, Croatia, May 19–22, 2014. 47

[5] P. E. Fayemi, K. Wanieck, C. Zollfrank, N. Maranzana, and A. Aoussat, Biomimetics: Process, tools and practice, *Bioinspiration and Biomimetics*, 12:011002, 2017. DOI: 10.1088/1748-3190/12/1/011002. 47, 49

[6] ISO 18458:2015, *Biomimetics—Terminology, Concepts and Methodology*, International Standards Organization, Geneva, Switzerland, 2015. https://www.iso.org/standard/62500.html DOI: 10.3403/30274979. 47

[7] T. Lenau, K. Helten, C. Hepperle, S. Schenkl, and U. Lindemann, Reducing consequences of car collision using inspiration from nature, *Proceedings of IASDR: 4th World Conference on Design Research*, Delft, The Netherlands, Oct. 31–Nov. 4, 2011. 47

[8] D. DeLuca, *The Power of the Biomimicry Design Spiral*, Biomimicry Institute, Missoula, MT, 2017. https://biomimicry.org/biomimicry-design-spiral/ 47

[9] N. Cross, *Engineering Design Methods—Strategies for Product Design*, Wiley, Chichester, UK, 2008. 48

[10] G. Pahl, W. Beitz, J. Feldhusen, and K.-H. Grote, *Engineering Design: A Systematic Approach*, 3rd ed., Springer, London, UK, 2007. DOI: 10.1007/978-1-84628-319-2. 48

[11] S. Keshwani, T. A. Lenau, S. Ahmed-Kristensen, and A. Chakrabarti, Comparing novelty of designs from biological-inspiration with those from brainstorming, *Journal of Engineering Design*, 28:654–680, 2017. DOI: 10.1080/09544828.2017.1393504. 48

[12] V. Srinivasan and A. Chakrabarti, Investigating novelty–outcome relationships in engineering design, *Artificial Intelligence for Engineering Design, Analysis and Manufacturing*, 24:161–178, 2010. DOI: 10.1017/s089006041000003x. 48

[13] M. Helms and A. K. Goel, The four-box method: Problem formulation and analogy evaluation in biologically inspired design, *Journal of Mechanical Design*, 136:111106, 2014. DOI: 10.1115/1.4028172. 50, 51

[14] Merriam-Webster, *Thesaurus*, Springfield, MA. https://www.merriam-webster.com/thesaurus 53

[15] Princeton University, *WordNet: A Lexical Database for English*, Princeton, NJ. https://wordnet.princeton.edu/ 53

[16] The Biomimicry Institute, *AskNature: Innovation Inspired by Nature*, Missoula, MT. https://asknature.org/ 54

[17] The Biomimicry Institute, *The Biomimicry Taxonomy*, Missoula, MT. https://asknature.org/resource/biomimicry-taxonomy/ 54

[18] Protected Planet, https://www.protectedplanet.net/en 55

[19] T. A. Lenau, Do biomimetic students think outside the box? *Proceedings of the 21st International Conference on Engineering Design (ICED17), Vol. 4: Design Methods and Tools*, 4:543–551, Vancouver, Canada, Aug. 21–25, 2017. 55

[20] BIOSIS Previews®. https://www.ebsco.com/products/research-databases/biosis-previews 56

[21] Encyclopedia of Life. https://eol.org/ 56

[22] T. A. Lenau, S. Keshwani, A. Chakrabarti and S. Ahmed-Kristensen, Biocards and level of abstraction, *Proceedings of the 20th International Conference on Engineering Design (ICED15)*, pages 177–186, Milan, Italy, July 27–30, 2015. 56

[23] Posters from DTU-BID course. http://polynet.dk/BID/ 57

[24] J. Davenport, R. N. Hughes, M. Shorten, and P. S. Larsen, Drag reduction by air release promotes fast ascent in jumping emperor penguins—a novel hypothesis, *Marine Ecology Progress Series*, 430:171–182, 2011. DOI: 10.3354/meps08868. 57

[25] E. Graeff, N. Maranzana, and A. Aoussat, Biomimetics, where are the biologists?, *Journal of Engineering Design*, 30:289–310, 2019. DOI: 10.1080/09544828.2019.1642462. 59

CHAPTER 6

Solution-Driven Biologically Inspired Design

> I think the biggest innovations of the 21st century will
> be at the intersection of biology and technology.
> A new era is beginning.
> STEVEN P. JOBS (2011)[1]

6.1 INTRODUCTION

Biologically inspired design (BID) can be approached from two distinctly different directions [1–3], leading to PROBLEM-DRIVEN BID and SOLUTION-DRIVEN BID. Whereas the former was described in Chapter 5, this chapter explains the latter approach which is called BIOLOGY-PUSH BIOMIMETICS by the International Standards Organization because it is the experience from biology that initiates and drives industrial application [4]. Although the term BOTTOM-UP BIONIK was initially used by researchers at the Technische Universität München [5], solution-driven BID is now referred to as SOLUTION-BASED BIOMIMETICS by them [6].

The challenge in solution-driven BID is to identify technical applications that will benefit from a set of solution principles identified from the bioworld. Solution-driven BID is often initiated by biologists with deep insights into biological functionalities but, typically, only little knowledge of technical applications and design methodologies. The search for applications followed by design work can therefore be quite arduous tasks for many biologists. Nevertheless, several examples of solution-driven BID exist in the literature, two of the most well-known examples originating from burdock seeds that inspired Velcro™ [7] and the self-cleaning leaves of the lotus plant [8] that inspired superhydrophobic surfaces [9, 10]. A few examples are described in this chapter to illustrate how observations of and inspirations from bioworld phenomena have been transformed into technical applications, followed by a description of the eight steps of an approach to implement solution-driven BID [11].

---

[1]Walter Isaacson, *Steve Jobs*, Simon & Schuster, New York, 2011.

6.2   EXAMPLES OF SOLUTION-DRIVEN BID

6.2.1   MYCELIUM BIO-COMPOSITES

Mycelium is the root system of mushrooms and other types of fungus. It is typically a fine mesh of tiny white strands referred to as hyphae [12]. The root system grows very rapidly through soil where it degrades dead lignocellulosic material such as straw and wood into nutrients used by the fungus. Other organisms also benefit from this process, since many fungi form symbiotic relationships with plants. The fungus lives at the base of many plants, the mycelium spreading along the plant's roots. In a symbiotic relationship, the plant supplies the fungus with carbon in the form of sugars made via photosynthesis in exchange for water and minerals such as phosphorus [13]. The exchange is actually more complex since the mycelium also serves as a connector between larger plants such as trees and small seedlings for exchange of water and nutrients.

The fungi, especially due to the mycelium, act as important waste-treatment actors in the bioworld, first degrading organic material and then transforming it into other types of organic material. This process can be technologically adapted for the production of MYCELIUM BIO-COMPOSITES that can be used for insulation, packaging material, and other lightweight structural products [12, 14, 15]. Agricultural waste streams comprise straw and husk which can be transformed into porous solids using fungi [16].

The left panel of Fig. 6.1 illustrates a corrugated panel made of a mycelium bio-composite. The surface is similar to that of plastics but is a bit rougher in texture and appearance. The natural origin of the bio-composite is evident to both eyes and fingers, promoting its use as a natural and biodegradable alternative to foamed plastics. The American company Ecovative has commercialized the manufacturing process for a range of foamy products [17, 18]. The first products were insulation and packaging items to replace foamed polystyrene. In these products, sometimes referred to as mycocomposites, the mycelium functions as a self-assembling biological binder for agricultural byproducts. Ecovative has also used the mycelium-based technology to produce a refined material for clothing fabrics and foamy skincare products.

As the mycelium is edible, mycelium bio-composites can be consumed as food. It is possible to achieve a texture and flavor similar to meat and in that way offer a vegetarian alternative. No animal products are used at all, which makes mycelium bio-composites attractive as food for vegans.

A limitation of the currently available mycelium bio-composites is their relatively high weight; hence, these materials cannot compete with the very lightweight foamed plastics. This has to do with the manufacturing method in which the finely chopped agricultural wastes are kept in shape by loading them into the cavity of a mold, thereby limiting the growth of the hyphae to the void regions between the fibers of the agricultural material. This was the experience of a design team at Danmarks Tekniske Universitet (DTU) when making the foam core of a 2-m-long surfboard of a mycelium bio-composite, shown in the right panel of Fig. 6.1. Although such a large object could be made with the required strength, it was still too heavy for the intended purpose.

Figure 6.1: (Left) Corrugated panel made of a mycelium bio-composite with a similar but more natural appearance compared to foamed plastics. (Right) Foam core of a 2-m-long surfboard made from hemp fibers bound together by mycelium. Courtesy: Dan Skovgaard Jensen, Kristian Ullum Kristensen, and Lasse Koefoed Sudergaard.

To improve the mycelium bio-composite, DTU researchers are working to combine the mycelium growing process with 3D printing [16]. One approach is to 3D print a porous matrix material in which the fungus grows much the same way as it does in the bioworld when degrading dead lignocellulosic material. Another approach is to use a 3D-printing technique in which the printing nozzle is maneuvered by a robotic arm to place the matrix material in space in the same way as spiders make their webs. After the hyphae spread in the 3D web, the resulting foamy material is very light and highly suitable for high-performance sandwich composites.

6.2.2   BOMBARDIER-BEETLE SPRAY

Ground beetles of many species such as *Stenaptinus insignis* [19] and *Brachinus crepitans* [20] are commonly called bombardier beetles because they exhibit an extraordinary self-protecting behavior. When approached by a predator such as an ant, a bombardier beetle sprays a boiling liquid toward the approaching predator, as depicted in Fig. 6.2. The liquid is ejected through a nozzle at the abdomen which can be directed to point toward the desired target. The amazing feature is that it is possible for the beetle to generate and handle a boiling liquid without harming itself. Another remarkable feature is the way in which the very hot aerosol is made. A gland containing hydrogen peroxide and another gland containing hydroquinone shoot their respective contents through the anus. When the two liquids mix with the enzymes catalase and peroxidase, hydrogen peroxide decomposes into water and oxygen and hydroquinone oxidizes

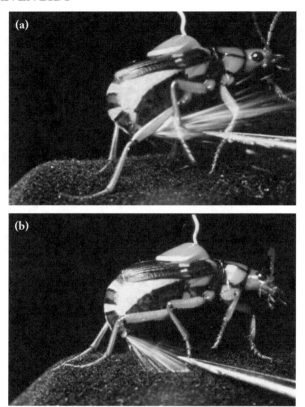

Figure 6.2: Hot spray is used by *Stenaptinus insignis* as a defense against predators [19]. Copyright (1999) National Academy of Sciences, U.S.A.

into p-quinones. Both reactions are exothermic, bringing the mixture to the boiling point and vaporizing it partially before expulsion along with free oxygen.

At the University of Leeds, the entire defense mechanism of the bombardier beetle species was found relevant to gas turbine igniters [20, 21]. The initial part of a research project undertaken at Leeds can be considered to be problem driven, as the desire to improve the combustion process in a gas turbine led to interest in a biological phenomenon. After studying the spray mechanism in the beetle, researchers constructed a scaled-up replica of the combustion chamber to demonstrate a similar spray formation. It was soon realized that the fascinating and remarkable properties of the bombardier spray mechanism could be useful for pharmaceutical sprays, fire extinguishers, and fuel injectors in combustion engines. That realization moved the work from problem-driven BID toward solution-driven BID. This can be seen as a definition of the attractive characteristics of a biological phenomenon (which is the first step in solution-driven

Figure 6.3: Tubercles on the leading edges of the flippers of a humpback whale improve lift and reduce drag as well as the risk of stalling. Courtesy: Whit Welles (Wwelles14) https://commons.wikimedia.org/w/index.php?curid=2821387.

BID, as explained in Section 6.3.2) for a spray technology that can reduce the environmental impact typical of existing spray technologies [21]. That understanding led to the identification of spray applications, such pharmaceutical sprays and fire extinguishers, that release polluting gases such as propane into the atmosphere.

The biomimetic spray technology is being applied in other scenarios too. For example, exhaust from internal combustion engines contains nitrogen oxides (NO_x), which contribute to smog and acid rain [22]. The release of NO_x is normally regulated by flow-restricting mixers. However, the principles for vapor formation in the bombardier beetles can be exploited to inject small droplets of a solution of urea into the exhaust and thereby inhibit NO_x release [23].

Swedish Biomimetics 3000 is commercializing the bombardier-beetle spray technology [24, 25]. This industrial company realized the potential of this biomimetic technology and began to explore applications in diverse industrial sectors, e.g., for air humidifiers in supermarkets.

6.2.3   TUBERCLES FOR FLOW CONTROL

Serendipity can play a big role in solution-driven BID. Frank Fish, a biology professor at West Chester University, happened to notice a curious feature in a figurine of three humpback whales (*Megaptera novaeangliae*) displayed at an antiques store [26]. The leading edges of the large flippers of the whales in the figurine were not straight but had tubercles. A little research showed that the artist had not made an error; indeed, the flippers of humpback whales have tubercles, as illustrated in Fig. 6.3.

Why do whale flippers have this strange geometry? Flippers are adaptations of hands. The biologist Fish found that the tubercles closely follow the joints in the phalanges of the "fingers" in each flipper. The flippers of a fully grown whale are about 3.60 m long, relatively long for an animal that is four times longer. The natural assumption for the biologist was that the tubercles serve a special purpose. This turned out to be true because humpback whales are very agile swimmers and can quickly change direction when swimming at a high speed. Unlike whales of other species, a group of humpback whales forms a circle when hunting a shoal of fish. The whales release bubbles from their blowholes to collectively form a cylindrical curtain to confine the shoal. The curtain is tightened as the radius of the circular formation decreases, delivering the prey in densely packed mouthfuls to the predators.

Computer models using the equations of fluid dynamics as well as experiments confirmed that the tubercles affect motion in a fluid significantly: lift is increased and drag is reduced [27]. These features explain the agility of humpback whales. The tubercles also reduce the risk of stalling which can happen when the lift of the flipper suddenly drops.

Having uncovered the physical principles underlying the fascinating capability of an animal arising from its anatomy, Fish wondered which technical applications could benefit from the enhanced flow characteristics when using a flipper or a fin with tubercles on the leading edge. This is the classic initiation of solution-driven BID, which justifies the appellation BIOLOGY-PUSH BIOMIMETICS. Many applications were investigated [28]. One is on the fin of a short surfboard which would enable a surfer to make a more sudden cutback, i.e., change direction when riding a wave. Another is on the keel of a sailing boat which can allow the boat to make tighter turns.

Like water, air is a fluid. Could applications in air also benefit from tubercles? Truck mirrors can be fitted with tubercles to reduce the drag and thereby improve fuel economy. The same effect can be achieved by adding tubercles to fins on racing cars. Helicopter rotors can deliver more lift with a reduced risk of stalling. Fans in stables can save energy while also becoming quieter. Windmills can generate more energy because of reduced drag.

Many good proposals for possible applications of the tubercles on the flippers of humpback whale emerged. Which of those proposals becomes commercial depends on the trade-off between achievable technical benefits and possible drawbacks as well as on the ease of production.

6.2.4   ABALONE-SHELL ARMOR

Abalone is the common name for marine mollusks belonging to the family Haliotidae. An abalone shell is shown in Fig. 6.4. The inner layer of the shell is made of nacre which is extremely tough, considering that most of it is aragonite which is very brittle because it is essentially chalk. The toughness can be measured as the specific work of fracture which is 0.2 J m^{-2} for aragonite but 400 J m^{-2} for nacre [29]. The explanation for nacre being 2,000 times tougher than aragonite is found in the layered "brick-and-mortar" micromorphology also shown in Fig. 6.4 [30, 31]. The bricks are plates of aragonite and the mortar is a ductile proteinaceous material [32].

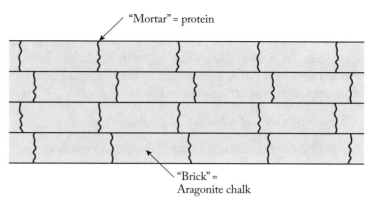

Figure 6.4: (Left) Nacre in an abalone shell. (Right) Schematic of the crack-resistant brick-and-mortar micromorphology of the abalone shell.

Toughness is often explained as prevention of crack propagation. A propagating crack is arrested when it encounters the proteinaceous mortar. When the shell experiences an impact from a crab or another predator, the impact energy causes the formation of microcracks in the aragonite plates. In a more homogeneous material, the microcracks would propagate and cause a failure, but the proteinaceous mortar absorbs the impact energy by deforming elastically and distributing part of the energy for microcrack formation in many other aragonite plates. Shell failure is thereby averted, the abalone shell thus providing an example of contraindicated performance.

The abalone shell happens to provide a documented example of a misapplication of solution-driven BID [1]. Fascinated by the impact resistance of the abalone shell, a group of engineering students at the Georgia Institute of Technology exploited the brick-and-mortar micromorphology for a bullet-proof vest. It was clearly a solution-driven approach where the inspiration came from a biological solution and was applied to a technical problem. However, the design team did not approach the exercise with sufficient rigor, going directly from a description of a fascinating functionality of a biological structure to a detailed specification of a solution to an appealing technical problem. They did not spend time on a closer analysis of what the properties of the abalone shell actually are and what type of impact it is best suited for. The abalone shell is very good at resisting the force from the jaws of a predator which typically applies the force at a slow speed. This is very different from the very sudden impact of a bullet flying at a high speed. Furthermore, the team designed the vest mimicking not only the micromorphology but also the chemical constituents (small flakes of chalk and elastic matrix) of the shell. Not only was the vest incapable of resisting bullets, it was much too heavy as well.

6.3   STEPS FOR SOLUTION-DRIVEN BID

Section 5.2 provides a five-phase implementation scheme along with several ways to adopt in each phase. In contrast, literature contains much less information on formal implementation of solution-driven BID. Researchers at the Georgia Institute of Technology have formulated a seven-step implementation plan as follows [1]:

  (i) become aware of a biological phenomenon,

 (ii) define the functionalities that brought attention to that biological phenomenon,

(iii) extract the key principles underlying the attractive functionalities,

(iv) specify the usefulness of the biological functionalities for human activities,

 (v) search for technical problems that can be solved using the identified functionalities,

(vi) select a technical problem from the ones identified, and

(vii) apply the key principles to that technical problem.

However, the instructions available in the design literature for some of these steps are scant. DTU researchers therefore developed an eight-step procedure to implement solution-driven BID, with inspiration from the way application search is done for conventional technology [11].

6.3.1   APPLICATION SEARCH

Application search is routinely carried out in any company that is focused on using a specific production technology. In order to ensure future sales, the company will regularly evaluate its present portfolio of products and search for new areas that will benefit from its production technology. The company will encounter challenges when seeking expansion into industrial sectors that it has no experience in. Due to this limitation, the company will serve as a subcontractor to companies that have both the required experience and contacts with end users.

Another limiting factor for such a company is that its principals are not trained in design thinking and are less experienced in working with open problems and large spaces of solutions. Instead, their forte is a deep knowledge of the specific production technology which enables their company to mass produce at a competitive price. The company will also be good at improving the technology to incorporate new features. But unlike companies with end-user contact (such as manufacturers of furniture or household appliances), it does not have a well-defined user group that can be explored to identify expansion potential. Identification of industrial sectors for expansion can therefore be a challenge. APPLICATION SEARCH is a way to meet this challenge.

As an example, consider application search carried out by a company that specializes in re-action molding of polyurethane, which is used to make toilet seats, panels for interior decoration, and dashboards of cars. A design-oriented approach to application search for this company is to

Figure 6.5: Pinart toy for children.

first identify the attractive characteristics of the reaction-molding technology and then search for end-user applications in order to identify candidate companies that will benefit from its technology. The low tooling price for manufacturing polyurethane objects enables: (i) the production of small batches of custom-designed objects, (ii) a high degree of freedom for free-form geometry, and (iii) the production of lightweight components with foamed core that can be inserted in metal, wood, and textile items. For each of these three enabling attributes, an open search for applications can be made, in brainstorming sessions and/or on internet search engines.

Another example is a project carried out by two engineering students to develop a new type of production technology based on the pinart toy shown in Fig. 6.5 [11]. The production technology is based on a mold that can change shape on demand and hence be useful for casting individually shaped items. An application search to justify the development of the mold identified 136 quite different applications encompassing prosthetics, contact lenses, hearing aids, chocolates, compact-disk covers, jewelry, propellers for sailing boats, and concrete bridges. A specific application must be selected in the development phase, since many parameters for the production tool (in this case, the mold), such as dimensions, accuracy, resolution, and throughput rate depend on the application. Based on an analysis of the applications and dialogue with possible collaborators for each of the application areas, two applications were selected: (i) a tool to fabricate individually shaped curved concrete facade elements and (ii) a tool for inscribing marks on casts to enable subsequent traceability during manufacture. The two resulting tools are shown in Fig. 6.6. Both applications are very different and addressed very different business areas.

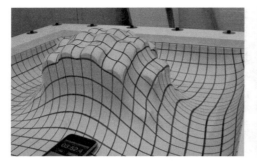

Figure 6.6: (Left) A tool for the fabrication of individually shaped curved concrete facade elements [33] and (right) a tool for inscribing marks on casts [34], both developed based on the pinart toy shown in Fig. 6.5.

Figure 6.7: Lotus leaves repel water and stay clean thereby.

6.3.2 EIGHT-STEP PROCEDURE

With the knowledge that application searches are routinely carried out in some industrial sectors, DTU researchers devised an eight-step procedure that has some overlap with the seven-step implementation plan from the Georgia Institute of Technology. The eight steps of the DTU approach for solution-driven BID are provided in Table 6.1.

The eight-step procedure is exemplified in Table 6.1 by the leaves of the lotus (*Nelumbo nucifera*), a plant native to many tropical countries. Considered sacred by Hindus, Buddhists, and Jains, lotus grow in wetlands, ponds, and lakes. The remarkable characteristic of this plant is that the ventral surfaces of its leaves stay clean even in dirty surroundings because those surfaces are superhydrophobic [9], as may be noticed in Fig. 6.7.

Table 6.1: The eight steps of the DTU approach for solution-driven BID along with the lotus-leaf example of self-cleaning surfaces in the bioworld

No.	Step	Lotus-Leaf Example
1	Define attractive characteristic of the biological phenomenon	*Characteristic:* Water repellence
2	Make an open search for applications	Self-cleaning vehicles
3	Formulate constraints to limit the scope of the search	*Constraint:* Only applications for which cleaning is difficult
4	Apply constraints one by one to eliminate some results of Step 2	Inside the shield of a lawn mower
5	Create a concept for each result of Step 4	Coat the inside of the shield of the lawn mower for cleaning with a garden water hose
6	Consult selected stakeholders	Talk to a few gardeners
7	Repeat Step 5 for new application identified during consultations	Wheelbarrows
8	Assess every concept against predefined criteria	*Criteria:* (i) Longer life time for lawn mower and (ii) lower risk of spreading pests

Step 1. Solution-driven BID begins with the awareness of a biological phenomenon that could either constitute or provide a solution to a technical problem that has not been identified. Thus, solution-driven BID can be initiated by merely an interest in an animal or a plant with a fascinating behavior or capability. It can also be initiated by a biologist who has studied biological organisms of a certain species or genus for many years and begins to wonder which engineering applications could benefit from the biological insight. Defining the biological solution then requires a description of its characteristics that may be relevant to some applications. Biological organs are typically multifunctional, so it may be arduous to describe all of its characteristics. Fortunately, a complete description is not called for, since it was a specific characteristic that drew attention. In the first step, that attractive characteristic of the biological phenomenon must be defined.

The persistent clean condition of lotus leaves can be explained by its water-repellence characteristic which prevents dust particles and other detritus from attaching to its ventral surface. The superhydrophobicity is responsible for the formation of water beads that roll off the surface, thereby removing foreign matter. In turn, this superhydrophobicity arises from surface topology at the 10-μm length scale [35]. However, as the matte appearance of lotus leaves is quite different from the glossy appearance of clean and hygienic surfaces, the superhydrophicity due to surface topology may not be attractive enough for certain applications.

Step 2. Next, an open search is made for applications that will benefit from the attractive characteristic defined in the first step. This can be done in different ways, but a simple one is for the design team to brainstorm in order to answer the following question: "In what situations can the described characteristic be advantageous?"

The question for the lotus-leaf example is: "Where can self cleaning be advantageous?" A more general question is: "In which situations do surfaces become dirty?" The unwanted consequence of having a matte surface could lead to the following question: "Where are clean but non-glossy surfaces required?"

Step 3. The characteristic defined in the first step will most likely result in finding a large number of possible applications in the second step. Therefore, the third step requires the formulation of constraints that will not only limit the scope of the search but also force deeper explorations of the fewer possible applications.

A constraint can require focus on items of specific types—e.g., household items, leisure and sports equipment, hospital articles, professional tools, etc. Another constraint can be on the type of materials deemed acceptable. A third way to approach setting up constraints could be to analyze daily or professional routines while looking for activities that benefit from the defined characteristic. Such a routine could be what a person does while working in an office or while traveling every week to meet clients on site. Professional routines can also be incorporated by choosing a professional activity such as gardening, hospital sanitation, painting houses, and graffiti removal. The framing of a context makes it easier to imagine where the defined characteristic of the biological solution may be beneficial.

A simple constraint for the lotus-leaf example is to focus on situations in which particle accumulation is undesirable and the particles are difficult to remove.

Step 4. Application of the constraints formulated in the third step will eliminate many of the possible applications identified in the second step. The constraints can be applied either sequentially or concurrently. Brainstorming by the design team will deliver context-specific applications.

For the lotus-leaf example, application may be sought for lawn mowers in which the operator is protected from the cutting blade by a shield. The cut grass often sticks to the inside surface of the shield and is not easy to remove. Another possible application is for a house painter's tools to have non-stick surfaces. Likewise, exterior walls of office buildings require treatment to prevent becoming canvases for graffiti artists.

Step 5. For each of the results of the constrained search undertaken in the fourth step, a concept has to be created. As explained in Section 2.4.4, whereas an idea is merely a principle for how to solve a problem, the application of that principle in a specific context leads to a CONCEPT because it satisfies the context-specific constraints. The intended performance of each concept must be described in concrete terms in the fifth step.

For the lotus-leaf example, a concept for the lawnmower is to endow the internal surface of the shield with topology at the 10-μm length scale to prevent wet cut grass from attaching to that surface. Likewise, a concept for the house painter's tools is have the exposed surface of every tool with a similar topology to prevent paint from adhering to the exposed surface. Finally, providing the surfaces of walls with a similar topology will deter graffiti artists.

Step 6. Each concept for every application has to be discussed with knowledgeable stakeholders in the sixth step. The stakeholders should be presented with the relevant concept(s) instead of being asked about possible applications. Some stakeholders are very likely to have reservations about why a concept may not work well in the real world, but the main point is to stimulate their creativity so they may come up with their own application proposals. Often it is easier to be creative when criticizing a concept.

For the lotus-leaf example, the stakeholders to be consulted should be gardeners for the lawn-mower concept, house painters for the painting-tools concept, and janitors for the graffiti-prevention concept.

In the case of the production technology based on the pinart toy shown in Fig. 6.5 [11], a concept was of a flexible mold for use by sandcasting companies. When sandcasting personnel were consulted on this concept, they informed the design team that the need for flexible molds is insignificant but a major need exists for traceability during the manufacturing process. If an individual code or number could be inscribed on each cast by the mold, then it would be possible to trace each cast subsequently. The quality of representative casts from a batch could be assessed and related to the personnel who produced that batch as well as to the specific material composition used. The company could in this way get a better quality-assurance system. The design team had not been aware of the need for traceability, but consultation with knowledgeable stakeholders led to a new application of their technology.

Step 7. The penultimate step is a repetition of the fifth step for the new applications identified by knowledgeable stakeholders during the sixth step.

For the lotus-leaf example, gardeners could suggest superhydrophobic surfaces for wheelbarrows, house painters could suggest similar surfaces for lunchboxes, and janitors for walls in children's bedrooms and school rooms.

Step 8. The final step of the DTU approach for solution-driven BID is to assess every concept with respect to a set of predefined criteria which could include the expected market capacity and societal impact.

6.4   REFERENCES

[1] M. Helms, S. S. Vattam, and A. K. Goel, Biologically inspired design: Process and products, *Design Studies*, 30:606–622, 2009. DOI: 10.1016/j.destud.2009.04.003. 61, 67, 68

[2] T. A. Lenau, A.-L. Metze, and T. Hesselberg, Paradigms for biologically inspired design, *Proceedings of SPIE*, 10593:1059302, 2018. DOI: 10.1117/12.2296560. 61

[3] L. H. Shu, K. Ueda, I. Chiu, and H. Cheong, Biologically inspired design, *CIRP Annals—Manufacturing Technology*, 60:673–693, 2011. DOI: 10.1016/j.cirp.2011.06.001. 61

[4] ISO 18458:2015, *Biomimetics—Terminology, Concepts and Methodology*, International Standards Organization, Geneva, Switzerland, 2015. https://www.iso.org/standard/62500.html DOI: 10.3403/30274979. 61

[5] T. Lenau, K. Helten, C. Hepperle, S. Schenkl, and U. Lindemann, Reducing consequences of car collision using inspiration from nature, *Proceedings of IASDR: 4th World Conference on Design Research*, Delft, The Netherlands, Oct. 31–Nov. 4, 2011. 61

[6] K. Wanieck, P.-E. Fayemi, N. Maranzana, C. Zollfrank, and S. Jacobs, Biomimetics and its tools, *Bioinspired, Biomimetic and Nanobiomaterials*, 6:53–66, 2017. DOI: 10.1680/jbibn.16.00010. 61

[7] S. D. Strauss, *The Big Idea: How Business Innovators Get Great Ideas to Market*, pages 14–18, Dearborn Trade Publishing, Chicago, IL, 2002. 61

[8] C. Neinhuis and W. Barthlott, Characterization and distribution of water-repellent, self-cleaning plant surfaces, *Annals of Botany*, 79:667–677, 1997. DOI: 10.1006/anbo.1997.0400. 61

[9] X.-M. Li, D. Reinhoudt, and M. Crego-Calama, What do we need for a superhydrophobic surface? A review on the recent progress in the preparation of superhydrophobic surfaces, *Chemical Society Reviews*, 36:1350–1368, 2007. DOI: 10.1039/b602486f. 61, 70

[10] J. Wang, L. Wang, N. Sun, R. Tierney, H. Li, M. Corsetti, L. Williams, P. K. Wong, and T.-S. Wong, Viscoelastic solid-repellent coatings for extreme water saving and global sanitation, *Nature Sustainability*, 2:1097–1105, 2019. DOI: 10.1038/s41893-019-0421-0. 61

[11] T. A. Lenau, Application search in solution-driven biologically inspired design, *Proceedings of the 22nd International Conference on Engineering Design (ICED19)*, pages 269–278, Delft, The Netherlands, Aug. 5–8, 2019. DOI: 10.1017/dsi.2019.30. 61, 68, 69, 73

[12] F. V. W. Appels, S. Camere, M. Montalti, E. Karana, K. M. B. Jansen, J. Dijksterhuis, P. Krijgsheld, and H. A. B. Wösten, Fabrication factors influencing mechanical, moisture- and water-related properties of mycelium-based composites, *Materials and Design*, 161:64–71, 2019. DOI: 10.1016/j.matdes.2018.11.027. 62

[13] J. D. Birch, S. W. Simard, K. J. Beiler, and J. Karst, Beyond seedlings: Ectomycorrhizal fungal networks and growth of mature, *Pseudotsuga menziesii, Journal of Ecology*, 2020. DOI: 10.1111/1365-2745.13507. 62

[14] F. V. W. Appels, *The use of fungal mycelium for the production of bio-based materials*, Ph.D. Dissertation, Utrecht University, The Netherlands, 2020. https://dspace.library.uu.nl/bitstream/handle/1874/390884/5e1c62cd1b0f1.pdf 62

[15] C. Bruscato, E. Malvessi, R. N. Brandalise, and M. Camassola, High performance of macrofungi in the production of mycelium-based biofoams using sawdust—sustainable technology for waste reduction, *Journal of Cleaner Production*, 234:225–232, 2019. DOI: 10.1016/j.jclepro.2019.06.150. 62

[16] O. Robertson, F. Høgdal, L. Mckay, and T. Lenau, Fungal future: A review of mycelium biocomposites as an ecological alternative insulation material, *Proceedings of NordDesign*, Kongens Lyngby, Denmark, Aug. 11–14, 2020. 62, 63

[17] E. Bayer and G. McIntyre, *Method for making dehydrated mycelium elements and product made thereby*, US Patent 2012/0270302 A1, October 25,1997. https://patents.google.com/patent/US20120270302A1/en 62

[18] Ecovative Design, *We Grow Materials*. https://ecovativedesign.com/ 62

[19] T. Eisner and D. J. Aneshansley, Spray aiming in the bombardier beetle: Photographic evidence, *Proceedings of U.S. National Academy of Sciences*, 96:9705–9709, 1999. DOI: 10.1073/pnas.96.17.9705. 63, 64

[20] N. Beheshti and A. C. McIntosh, A biomimetic study of the explosive discharge of the bombardier beetle, *International Journal of Design and Nature*, 1:61–69, 2007. DOI: 10.2495/d&n-v1-n1-61-69. 63, 64

[21] A. C. McIntosh, Biomimetic inspiration from fire and combustion in nature including the bombardier beetle, *Proceedings of SPIE*, 7401:74010F, 2009. DOI: 10.1117/12.825477. 64, 65

[22] D. T. Allen and D. R. Shonnard, *Sustainable Engineering: Concepts, Design, and Case Studies*, Prentice Hall, Upper Saddle River, NJ, 2012. 65

[23] P. Larsson, W. Lennard, O. Andersson, and P. Tunestal, A droplet size investigation and comparison using a novel biomimetic flash-boiling injector for AdBlue injections, *SAE Technical Paper* 2016-01-2211, 2016. DOI: 10.4271/2016-01-2211. 65

[24] Swedish Biomimetics 3000, *μLOT® technology*. https://sb3000.tech/ulot-process/ 65

[25] L.-U. Larsson, Swedish Biomimetics 3000®, *Bioinspired!*, 6(1):6–7, 2008. https://bioinspired.sinet.ca/content/february-2008-newsletter-issue-61 65

[26] A. S. Brown, From whales to fans, *Mechanical Engineering*, 133(5):24–29, 2011. DOI: 10.1115/1.2011-may-1. 65

[27] D. S. Miklosovic, M. M. Murray, L. E. Howle, and F. E. Fish, Leading-edge tubercles delay stall on humpback whale (*Megaptera novaeangliae*) flippers, *Physics of Fluids*, 16:L39–L42, 2004. DOI: 10.1063/1.1688341. 66

[28] F. E. Fish, P. W. Weber, M. M. Murray, and L. E. Howle, The tubercles on humpback whales' flippers: Application of bio-inspired technology. *Integrative and Comparative Biology*, 51:203–213, 2011. DOI: 10.1093/icb/icr016. 66

[29] A. P. Jackson, J. F. V. Vincent, and R. M. Turner, Comparison of nacre with other ceramic composites, *Journal of Materials Science*, 25:3173–3178, 1990. DOI: 10.1007/bf00587670. 66

[30] T. A. Lenau and T. Hesselberg, Biomimetic self-organization and self-healing, *Engineered Biomimicry*, A. Lakhtakia and R. J. Martín-Palma, Eds., pages 333–358, Elsevier, Waltham, MA, 2013. DOI: 10.1016/c2011-0-06814-x. 66

[31] M. Mirkhalaf, D. Zhu, and F. Barthelat, Biomimetic hard materials, *Engineered Biomimicry*, A. Lakhtakia and R. J. Martín-Palma, Eds., pages 59–79, Elsevier, Waltham, MA, 2013. DOI: 10.1016/c2011-0-06814-x. 66

[32] L. Addadi, D. Joester, F. Nudelman, and S. Weiner, Mollusk shell formation: A source of new concepts for understanding biomineralization processes, *Chemistry: A. European Journal*, 12:980–987, 2006. DOI: 10.1002/chem.200500980. 66

[33] T. H. Pedersen and T. A. Lenau, Variable geometry casting of concrete elements using pin-type tooling, *Journal of Manufacturing Science and Engineering*, 132:061015, 2010. DOI: 10.1115/1.4003122. 70

[34] N. K. Vedel-Smith and T. A. Lenau, Casting traceability with direct part marking using reconfigurable pin-type tooling based on paraffin-graphite actuators, *Journal of Manufacturing Systems*, 31:113–120, 2012. DOI: 10.1016/j.jmsy.2011.12.001. 70

[35] L. Gao and T. J. McCarthy, Wetting 101, *Langmuir*, 25:14105–14115, 2009. DOI: 10.1021/la902206c. 71

CHAPTER 7

Biologically Inspired Design for the Environment

> The earth, the air, the land and the water are not an inheritance from our forefathers but on loan from our children. So we have to handover to them at least as it was handed over to us.
> MOHANDAS K. GANDHI[1]

7.1 SUSTAINABILITY AND THE ENVIRONMENT

Concern about sustainable development is mounting as the number of people on our planet increases. In 1987 the Brundtland Commission of the United Nations [1, 2] defined SUSTAINABLE DEVELOPMENT as "development that meets the needs of the present without compromising the ability of future generations to meet their own needs." The commission considered three areas of concern for sustainable development: (i) the environment, (ii) social organization, and (iii) economy. Technological development as well the global organization of human society currently require the imposition of serious curbs on consumption, especially considering the limited ability of the biosphere to absorb diverse types of waste excreted by human activities. However, both technological development and social organization can be managed and improved to make way for enhanced economic growth and poverty removal. Sustainable development requires that the more affluent humans adopt lifestyles that are consistent with the planet's ecological well being—for instance, in their consumption of energy. Also, population increase needs to be in harmony with the changing productive potential of the ecosystem.

The quest for sustainable development was taken further in the 2030 Agenda for Sustainable Development which, in 2015, resulted in the United Nations General Assembly adopting 17 sustainable development goals (SDGs) [3]. The SDGs are operational goals focused on concrete actions. Figure 7.1 classifies all 17 SDGs in relation to the previously mentioned three areas of concern: the environment (also referred to as the biosphere), social organization, and economy [4].

---

[1]https://bestquotes.wordpress.com/2007/03/24/hello-world/

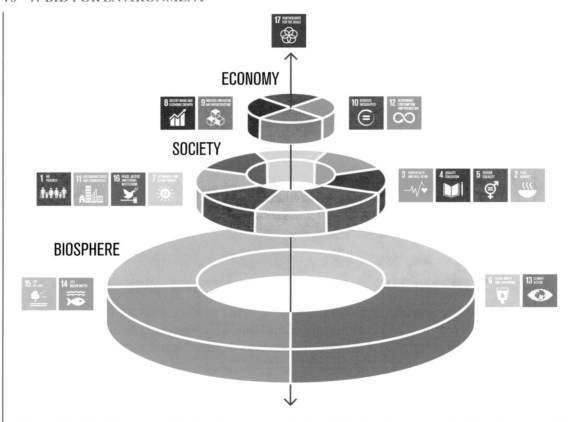

Figure 7.1: The 17 sustainable development goals classified for relevance to the biosphere, social organization, and economy. Credit: Azote Images for Stockholm Resilience Center, Stockholm University.

Biomimicry can help in addressing current actions and proposing new actions within all three areas, with focus on using inspiration from the bioworld to solve problems relating to the biosphere. The following SDGs can be impacted by biomimicry:

SDG 6: clean water and sanitation,

SDG 7: affordable and clean energy,

SDG 13: climate action,

SDG 14: life below water, and

SDG 15: life on land.

Both economy and social organization are human constructs and, even though inspirations for their improvement can be found in the bioworld, the dominant application of biomimicry is for technological solutions in line with SDGs 6, 7, 13, 14, and 15. The bioworld presents many avenues that can be adapted for circular economy, resource efficiency, and ecosystem balances.

7.2   MATTER OF SCALE

A crucial challenge to sustainable development is posed by the growing number of people on the planet. More people share the limited resources available, more people produce waste, and more people pollute. Equal opportunities for everyone being a human right, countries should aim at providing the highest standard of living consistent with the overall health of the biosphere that includes not only all humans but all other living organisms too.

An activity that seems to be only a small problem when carried out by a few people can turn out to be a huge problem when carried out by many people. Numerous examples show how small problems grow out of proportion when scaled to larger populations. In Jakarta, Indonesia, it is common practice for landowners to pump water from aquifers deep underground since piped water is not reliably available. However, with 10 million inhabitants this practice has caused land subsidence of as much as 4 m in the coastal areas of the city, thereby making it highly vulnerable to flooding [5]. Another example is eutrophication of lakes and rivers [6]. The use of fertilizers is desirable to increase agricultural yields, but the right amounts may not be applied at the correct times. Excess fertilizer will run off during rain and/or irrigation to cause increased growth of algae in rivers and lakes, leading to oxygen depletion and fish deaths on a large scale. This would not be a problem if confined to a few locations. However, widespread use of excess fertilizers impacts not only water bodies in landmasses but also causes dead zones in seas and oceans. Thus, a supposedly harmless action when undertaken by an individual can have grave repercussions on a large population in a large area when that same action is simultaneously implemented by more than a few individuals.

The *global environmental impact* (GEI) can be quantified as the product of three factors [7]: the number N of people on our planet, the per-capita economic activity E, and the *eco-efficiency* F defined as the environmental impact per economic activity. That is,

$$GEI = N \times F \times E.$$

The global population N in 2019 is around 7.5 billion, rising from 4 billion in 1974 and projected to rise to 10 billion in 2057 [8]. Concurrently, living standards (i.e., E) have improved for many people. In 1990, 36% of the global population was living in extreme poverty [9], defined by the World Bank as an income of US$ 1.9 a day [10]. Extreme poverty was reduced to 8% of the world population in 2018, which illustrates the fast pace at which the standard of life is being enhanced globally. To maintain an unchanged GEI, the eco-efficiency F must be decreased, i.e., the environmental impact for the economic activity must be lowered.

U.S., Canada, most European countries, Japan, Taiwan, South Korea, Saudi Arabia, Qatar, Bahrain, and, increasingly, parts of China and India have an opportunity in being role models for sustainable life-styles of desirable quality. These regions can demonstrate that it is possible to maintain a high standard of living that is consistent with sustainable development—a win-win situation. Sustainable development does not overburden planetary resources, and a high quality of life allows the citizen to reap the benefits of techoscientific advances.

A major requirement to engender this win-win situation is the low-cost production of energy from nonpolluting sources for millions of years to come. These sources include the sun, winds, tides, and reservoirs. Another major requirement is the minimization of the extraction of ores, minerals, and petroleum from the planetary crust by industrywide recycling of materials that have already been extracted. Improved quality of life for a growing population is possible only if both resource consumption and waste production are greatly reduced, resulting in improved eco-efficiency. Highly efficient systems in the bioworld can inspire technological developments for an effective transition toward sustainable development.

7.3   SUSTAINABLE PRACTICES FROM NATURE

The biosphere and the sun together can act as an ecosystem to sustain our species and countless others for very long periods of time compared to the human lifespan. Photosynthesis transforms solar energy into organic matter that is the basis of the food chain for other living organisms. The organic matter consists mainly of a few elements—carbon, oxygen, hydrogen, nitrogen, phosphorus, and sulfur—which can be combined into a large number of materials that later can be decomposed to form new materials. All materials are synthesized by self assembly into living organisms at ambient temperatures, and those organisms either excrete or decompose into materials that can be re-synthesized into living organisms. There is no external agent following some masterplan that determines the sequence and method of assembly. Instead, the entire process is embedded within the organisms so they can self replicate.

A good example of how nature produces materials with remarkable properties at ambient temperatures is the iridescent nacre found in mollusks. The material is very stiff and its hardness equals that of manufactured materials such as ceramics which require very high production temperatures. Located on the inside of the shell, nacre is a composite layered structure consisting of aragonite crystals (calcium carbonate) separated by very thin layers of a cross-linked protein. The structure is often referred to as brick-and-mortar structure due to its visual similarity to building materials [11].

The material-shaping mechanism is not completely understood but a good model has been described by Addadi et al. [12]. The layered structure is made through a long sequence of steps. The epithelial cells in the mollusk's mantle secrete the highly cross-linked periostracum protein layer under which a gel-like matrix is formed. The matrix includes both hydrophobic and hydrophilic proteins as well as colloidal particles of the chemically unstable amorphous calcium carbonate. Aragonite crystals form at nucleation sites and grow until the periostracum layer is reached. In between the crystals, chitin molecules are trapped so that the brick-and-mortar structure is formed.

An example from nature, but not the bioworld, is the way sediments are transported by sea currents along coasts. Soil is removed from places along coastlines, thereby causing land to disappear while cliffs are formed. The soil is moved by the sea currents and deposited at other places, typically forming headlands. Amazingly thus, water breaks down solid material and carries it over long distances simply through the persistent application of fairly small forces over long periods of time.

A similar mechanism is applied by humans to convert seabed into agricultural land. In Denmark, Germany, and the Netherlands, the tides are fairly large and the coastal areas are usually wide and flat. By deliberately placing obstacles to delay water brought over by the tides, the deposition of sediments is promoted causing the land to form above the sea level. This approach of using many repeated actions to create a shape abounds in nature. Each individual action is in itself not very powerful. In contrast, humans typically apply a lot of force for a short while to create similar formations—e.g., when an excavator is used to move soil.

The bioworld thus presents a very different approach for manufacturing materials compared to the approaches humans take. When considering length scales of up to 1 m, humans manufacture objects primarily by using high levels of energy and through a planned selection of materials [13]. For instance, polymers are typically processed by pressing the melted polymer into a mould while applying large forces. The bonding energies for polymerization are quite similar in magnitude for a large variety of polymers, whether manufactured in a factory or in the bioworld. But, biological systems do not use elevated temperatures and rely instead on chemical reactions when building blocks of the right basic materials are brought into position. Biological polymers are mainly proteins and polysaccharides in fibrous form, found in collagen, silk, muscles, and arthropod exoskeletons. Hard tissue in biology is mostly made from calcium and silicon with smaller fractions of iron, zinc, and manganese—all processed at ambient temperature [14].

7.4   CIRCULAR ECONOMY OF MATERIALS

Most biological materials can be used directly or indirectly by other organisms. Many mammals, for instance, eat the placenta after the birth of an offspring. All spiders produce silk but not all spiders spin webs. Webmaking saves a spider the energy-consuming effort of hunting by rapid locomotion, but it requires a sizable investment of proteins that the web is made of [15]. Many spiders eat their old webs so that the proteins are recycled to make new webs [16, 17].

Less directly, biological materials are broken down to simpler molecules by bacteria, making them useful for other organisms. Biodegradation is a well-known process whereby bacteria in the ground decompose dead organic material into carbon dioxide, nitrogenous compounds, and other materials [18]. Mycelia from fungi break down lignin and cellulose from plants. Fungi grow on dead trees on the forest floor after the wood has been moisturized, and the same can be seen in buildings with wooden structures. Thus, moisturized wood provides a good environment for fungi to grow and eventually break down the lignin and cellulose in the wood [18].

Colors in plants are usually produced using pigments [19] though sometimes structural colors are also found [20]. A structural color arises due to spectrally selective scattering of visible light in response to the morphology of a physical structure [21, 22]. Usually, the morphology has a repeat pattern that is tuned to a certain color. Whether dull or brilliant, a structural color is not produced by pigments, which is immensely important for biologically inspired design for environment in that material diversity is not enhanced by incorporating a structurally colored object.

Multifunctionality is commonplace in living organisms [23, 24], because fewer organs need to be formed, housed, and coordinated if those organs are multifunctional. As an example, a mouth is used for ingesting nutrients, releasing sounds, breathing, and showing affection. A multifunctional module can be incorporated in a variety of products, thereby reducing inventory costs, enhancing repairability, extending product lifetimes, and promoting standardization. Lifetime extension slows down the depletion of raw materials, reduces the consumption of energy for manufacturing, and reduces the volume of waste for disposal.

7.5   MUTUALLY BENEFICIAL COEXISTENCE

No organism in the bioworld exists on its own but is dependent on interactions with other organisms, whether of the same species or not. Within a species, wolves and dingoes hunt in coordinated groups for greater success, starlings fly in coordinated murmurations to confuse predators such as falcons, and fish similarly form schools (not to be confused with the much less coordinated shoals) to elude predators. Mammal mothers rely on kin to bring food and even look after infant offsprings.

Mammals rely heavily on symbiosis with microorganisms in their digestive tract. On average, a human has 0.2 kg of bacteria primarily in the intestines [25], not only to help break down food into substances that can be adsorbed through the intestine wall but also to supply signaling compounds essential to the mental health of the person [26]. Transplants of fecal matter can improve the health of humans suffering from a range of diseases [27].

Plants can produce carbohydrate building blocks through photosynthesis by extracting carbon from the air and water from the ground. However, they cannot extract minerals such as phosphorus from the soil and therefore benefit from a symbiotic relationship called mycorrhiza between their roots and mushrooms [18]. In exchange, the mushrooms get carbohydrates. Similarly, some bacteria extract nitrogen from air and supply it to plants as ammonia [18]. Nitrogen fixation is essential for the biosynthesis of amino acids, proteins, nucleic acids, and other nirogenous compounds.

Many animal species rely on social relationships to thrive and even exist. These relationships are very pronounced in social insects such as bees, ants, and termites. They are characterized by a division of labor whereby some individuals provide food, others nurture the eggs and larvae, and still others build and maintain the physical living facility. The individuals communicate using a range of signals including visual (e.g., color in flowers and waggle dance among bees), olfactory (e.g., pheromone trails made by ants) and acoustic (e.g., bees buzz with their wings). The role of the individual appears to be centrally controlled only to a limited degree, with guards allowing entry only to the inhabitants of the hive or pit. So, how do individuals know their roles and how to perform tasks without feedback from a central authority?

A very subtle type of communication to control flock behavior involves pheromones. A pheromone is an olfactory agent that, unlike many fragrances that animals consciously recognize, makes a short cut to the brain and produces almost instantaneous recognition. Pheromones assist in a range of different activities such as initiating alarms, attracting mates, and marking trails to be followed by others [28].

Inter-species communication is also commonplace. The approach of a fearsome predator leads to a single alarm signal that warns birds and mammals of diverse species to take evasive action [29]. Not only animals but plants also communicate. The roots of grasses and cereals of many types excrete chemical compounds that are processed by other plants to determine if the secreting plants belong to their family. This phenomenon has been deduced from the way in which the growth of roots of a certain plant is influenced by the roots of neighboring plants [30].

Human society too can benefit from symbiosis whereby the residual energy and materials from one company become resources for another company. Industrial symbiosis is an element in the circular economy which, apart from better utilization of resources, benefits society by increasing the number of jobs and boosting the Gross Domestic Product [31]. In the city of Kalundborg in Denmark, 11 public and private companies have formed a partnership facilitating a circular approach for the refinement of crude oil; production of insulin, fertilizers, and gypsum wallboards; and heating residences and office buildings [32]. The symbiotic activities direct waste energy, water, and materials from one company to another. For example, the insulin factory uses fermented sugars resulting in residual yeast biomass which is directed to a factory for producing fertilizers and biogas, the biogas is used in the gypsum factory for heating, and the residual thermal energy is transferred to a central heating plant. The result is better utilization of resources and materials combined with enhanced economy and employment.

Another example is the Danish Pig-City project that aims at combining different types of agri-businesses [33]. The project combines the husbandry of pigs and production of tomatoes with a slaughter house, an energy generation plant, and a bio-refinery. Heat from the piggery on the ground level is used for growing tomatoes in a greenhouse on the floor over the piggery. Organic waste from both the piggery and the greenhouse is treated in the bio-refinery to produce biogas for heating and fertilizers for the greenhouse.

7.6   ENERGY EFFICIENCY

Access to enough energy is a limiting factor for all physical and chemical processes in the bioworld. Just as for aeroplanes and helicopters, the range and duration of avian flight depend on how much energy does a bird have when it takes off into the air. Avian bodies have therefore evolved to have lightweight structures. Many large birds such as albatrosses, condors, and eagles exploit the warmer air currents for lift and thus minimize energy consumption by their pectoral and supracoracoideus muscles [34, 35]. Mammals regain energy cyclically when running. On the downstroke of a leg, the tendons, ligaments, and muscles stretch to store energy that is released on the offset. This is true for most animals, but a surprising phenomenon is seen for kangaroos which are very efficient energy regainers. At moderate speeds they are more energy efficient in terms of oxygen consumption compared to running bipeds and quadrupeds of similar size [36].

The force that impedes forward motion in a fluid is called drag. Several species have intricate mechanisms for reducing drag. Sharks are covered with tiny corrugated scales which introduce microturbulence close to the body surface. The microturbulence allows for a more laminar flow of seawater, thereby reducing the overall drag. The phenomenon has been mimicked in polymer films applied on aircraft to reduce drag [37]. The sharkskin scales are multifunctional since their corrugated shape also prevents fouling [38], because barnacles are not able to get a good grip and therefore fail to attach. Penguins reduce drag by releasing microbubbles of air trapped under their feathers. If necessary, a penguin can thus increase its speed several times over short distances, e.g., when chased by a predator [39].

7.7   DESIGN APPROACHES

Several approaches have been devised to support the designer toward the goal of sustainability enhancement. Formal guidelines help keep a tight focus toward that goal. The system-oriented approach of CIRCULAR DESIGN orients the designer not solely toward the manufacture of a specific product, but on its entire lifecycle to encompass raw materials, the use phase, and the utilization of waste products. A third approach is to assess the environmental footprint of the product.

7.7.1   ENVIRONMENTAL GUIDELINES

An approach suitable for the early-design stages when many product details are yet unknown is to use Green Design Guidelines (GDG) [40]. The widely used GDGs may have either very concrete forms such as the specification of acceptable materials, or be abstract by exhorting the embrace of techniques that produce less waste than those techniques that require remedial cleanup of the produced waste.

Another approach involves a systematic methodology to aim for efficiency in the use of energy and materials [41]. This approach comprises different types of efficiency (such as mechanical, material, and thermal efficiencies) and a framework to use bioinspiration. Once a type of efficiency is selected, analogies from the bioworld can help the designer by providing insight into functioning and efficient solutions.

The International Standards Organization defines biomimicry as "philosophy and interdisciplinary design approaches taking nature as a model to meet the challenges of sustainable development (social, environmental, and economic)" [42]. A distinction has been made in Chapter 1 between BIOMIMICRY and ENGINEERED BIOMIMICRY, the former being contained in the latter. Whereas engineered biomimicry does not need to be focused on reaching for sustainability goals, the term biomimicry—often associated with the Biomimicry Institute, an American non-profit organization—is focused on using inspiration from the bioworld to design solutions that contribute to sustainable development.

The Biomimicry Institute has a sister organization called Biomimicry 3.8 which is a consultancy working together with companies to solve design problems. One of the founding members of both organizations is Janine Benyus. The two organizations have developed a basic framework for design work [43] and the database Asknature [44] which allow searches for biological strategies to solve specific functional challenges. To support sustainable development, the organizations have formulated the following six lessons from the bioworld:

- evolve to survive,

- adapt to changing conditions,

- be locally attuned and responsive,

- use life-friendly chemistry,

- be resource efficient, and

- integrate development with growth.

Each lesson leads to specific guidelines, such as *incorporate diversity* and *use low-energy processes*, that are mainly concerned with the environmental part of sustainable development. These guidelines function in the same way as the criteria for evaluation of design proposals described in Chapter 2. When two proposed solutions are compared, the preferred one has to satisfy more guidelines in the best way. Thus, these guidelines are not absolute but indicate desirable outcomes.

In a study of biomimicry practices in the Nordic countries, it was found that only a few companies have combined biologically inspired design and environmentally conscious design [45]. But there are many examples of companies adopting either biologically inspired design or environmentally conscious design, so that their amalgamation is a realistic goal. In another study, designers were found to use several different sustainability frameworks when working with bio-inspiration, but without an established system of accountability [46].

7.7.2   CIRCULAR DESIGN

Designing a product with circularity in mind entails an insurance that recycled materials are used for production and that the product at the end of its life can be reused or recycled.

Circular economy is an approach to promote sustainable development with parallels to how resources are circulated in the bioworld. The Ellen MacArthur Foundation defines circular economy as an industrial economy that is restorative by intention [47]. Motivated by lessons learned from studies of living nonlinear systems, circularity is premised on the use of renewable energy, minimum consumption of chemicals, and eradication of waste. Circularity aims to optimize systems rather than their components. This is done by managing the flows of materials of two types: (i) biological nutrients that re-enter the biosphere safely and (ii) technical nutrients designed to circulate without diminishing in quality and without entering the biosphere.

Consequently, circular economy distinguishes between consumption and use of materials. It promotes a functional service model whereby the ownership of a product is retained with the manufacturer who acts as a service provider rather than as a product seller. The manufacturer therefore does not promote one-way consumption but ensures that the product will be reabsorbed in the economy after the end of its life.

Circularity can be applied to all types of industrial production. An example is the clothing industry. The current system is regarded as extremely wasteful and polluting from the initial production of textile fibers, through the production of fabrics and a wearable followed by repeated washes during use to the final after-use destiny of the wearable [48]. Typically, an item of clothing is discarded after the wearer is no longer interested in wearing it, although sometimes it can be passed on to another person. A cotton wearable may be collected by rag pickers as a raw material for producing paper and industrial wiping rags, there is hope that blended polymer/cotton wearables could be reprocessed after recovering and separating fibers of different materials, wool extracted from woolen wearables can be used for insulation panels for housing, acrylics and nylons can be reprocessed into blankets, but polyester wearables are mostly incinerated [49]. Circular economy in the clothing industry would be greatly facilitated by fiber-to-fiber recycling.

Cradle-to-cradle is an approach to maximize the positive effect of human activities on the environment as opposed to eco-efficiency that focuses on reducing damage to the environment [50, 51]. It is based on three key principles:

- waste equals food,

- use only energy provided currently by the sun, and

- celebrate diversity.

The first principle is inspired by the nutrient cycles seen in the bioworld. Instead of reducing waste, only that waste should be produced which another process can use as an input. The second principle dictates that all energy should come from the sun, i.e., from photovoltaic solar cells, solar thermal heaters, wind turbines, hydroelectric generators, and biomass incinerators. The third principle encourages design that respects local cultures and environments and also recognizes that nonhuman species have the right to thrive in their own ecosystems. A criticism of the cradle-to-cradle approach is that is does not address trade-offs between energy use and resource conservation, because even healthy emissions can adversely affect the ecosystem [50].

7.7.3   IMPACT ASSESSMENT

Life-cycle analysis is an approach to assess the eco-efficiency of a design. A comprehensive inventory is made of materials, energy, and chemicals used to make, distribute, use, and dispose of the product. The impacts of the materials, energy, and chemicals on the environment are also cataloged. In order to compare the eco-efficiencies of two different designs, a functional unit is defined to represent the desired functional performance. As an example, the functional unit can be used to facilitate the comparison of the eco-efficiencies of different ways of maintaining a golf course. A functional unit could be defined as the acreage of a certain terrain in which the height of grass must be maintained, which makes it possible to compare different methods to maintain grass height—e.g., using lawn mowers or letting a ruminant species such as goats or sheep graze.

Assessing the environmental impact is a fairly complex task since a design can have environmental effects through several mechanisms such as the emission of greenhouse gases leading to global warming, the emission of chlorofluorocarbons and halons leading to ozone-hole formation, and the acidification of lakes and rivers. When designing products, a simpler and less precise method is often used—namely, the use of indicators such as CO_2-equivalents. The indicators make it possible to compare quite different designs. For example, they can be used to compare the production of vegetables in heated greenhouses in a cold region with the production of vegetables produced in a warm region followed by transportation to the same cold region.

The use of life-cycle analyses has been criticized for not including the full potential of approaches such as biomimicry and cradle-to-cradle [50, 52]. Instead, a life-cycle analysis can become so easily focused on the function of a specific product that its goal can be best characterized as the *reduction of unsustainability*. Formulation of the functional unit can in some cases lead to ignoring ancillary issues whose consideration could have enhanced sustainability. Thus, a life-cycle analysis can lessen the use of energy and materials in a factory, but it will not address the improvement of air quality which could be very important for public health.

The life-cycle analysis of a product can be supplemented with clear criteria of when a product can be considered sustainable and when not. This is not an easy task, but attempts are in progress to define green products as having *zero waste*, producing *zero emissions*, and being *environmentally safe*.

7.8   GRAFTING "BIOLOGICALLY INSPIRED DESIGN" ONTO "DESIGN FOR ENVIRONMENT"

DESIGN FOR ENVIRONMENT aims at developing products to enhance sustainability without compromising functionality, cost, quality, etc. The bioworld presents many approaches that can be adapted for circular economy, resource efficiency, and ecosystem balance. As an example, microscopic scales on sharkskin swimsuits indeed reduce drag; likewise, sharkskin polymer films on aircraft and ships lower energy consumption [37]. But care must be exercised when transferring solution principles from the bioworld to industrial activities [53]. A bioworld phenomenon may appear simple at first glance but it may actually involve many intricate mechanisms to assure a desirable outcome. Its complexity may be inimical for adoption by designers. Additionally, a bioinspired solution may not comply with our ethics; for instance, the predator-prey relationship [54] is highly undesirable as a model for controlling the human population. Finally, an attractive solution principle may simply be impractical for adoption. As an example, a penguin can increase its speed several times over short distances underwater by releasing microbubbles of air trapped under its feathers [39], but the application of the same mechanism to reduce drag on a regular ship appears practically unimplementable.

The grafting of biologically inspired design onto design for environment requires a careful delineation of the design object. Design for environment is often focused on reducing the overall environmental impact of a specific product. An automobile engine that consumes less gasoline than its competitors delivering the same performance and driver satisfaction will comply with the objectives of design for environment. In other cases, a system involving many products and processes has to be considered. An example is the introduction of electric vehicles or hydrogen-powered vehicles that will necessitate the development of a comprehensive new infrastructure. In the bioworld, any organism relies on being part of a larger system comprising organisms of the same and different species. Environmental sustainability must therefore be addressed at both the product level and the system level, when a bioinspired solution principle has to be considered for adoption. The mutualistic relationship between plants, rhizobial bacteria, and mycorrhizal fungi which benefit from an exchange of nutrients and energy [18] illustrates how it can be insufficient only to consider an isolated object as the design object.

The design of a product or system typically involves the following four phases [55] described in detail in Chapter 2:

- definition and clarification of the need for the product or system (Sections 2.4.1–2.4.3),

- conceptualization of the product or system and the production/realization process (Sections 2.4.4–2.4.5),

- preparation of its embodiment to focus the attentions of all stakeholders (Section 2.4.6), and

- creation of the necessary detail for production and realization (Section 2.4.6).

Of these four phases, the conceptualization phase offers the most opportunities for implementing strategies associated with design for environment. These strategies include: reduction of material diversity, ease of disassembly and repairability for longer useful life, use of recyclable and recycled components, reduced use of toxic materials and nonrenewable resources, and ease of disassembly for circularity and recyclability.

An ever-growing compendium of bioinspired solution principles needs to be established for each of these strategies. This compendium could lead to the identification of new generic design principles for disruptive innovation. For example, egg shells and sea shells illustrate how chalk, a soft material, can be microstructured to bear huge static and dynamic loads. Thus, inferior materials can be biomimetically reformulated to deliver superior performance. The compendium would also promote multifunctionality, as exemplified by avian plumage being used for flight without significant increase of weight, water repellency, and conservation of body heat.

Design for environment brings additional constraints for biologically inspired design, which may considerably minimize the solution space. However, a clear environmental goal will facilitate a more focused search in the compendium and would stimulate creativity in finding new solutions. As an example, the nests of most birds are made from waste materials held together with friction and thus exemplify temporary structures that require very low investment but fulfill short-term needs for temporary housing.

The grafting of biologically inspired design onto design for environment will bring certain challenges. The evaluation of a radical solution from the bioworld may be difficult not only due to lack of data but also because of uncertainty in how it will affect use patterns and impact associated products. For example: inspired by the way spiders eat their own web every second day in order to regenerate the proteins [16, 17], a solution could be the local reuse of building materials. However, this will impact the business system for building materials and the working procedures of the construction industry. The uncertainty may be especially high when the context and the expected-use scenario for a product or system are not yet defined.

In summary, well-established theories and tools exists to analyze environmental impact and design to enhance sustainability. Still, design for environment can benefit from biologically inspired design to create novel solutions. For their integration into BIOLOGICALLY INSPIRED DESIGN FOR ENVIRONMENT, successful case stories and an ever-growing compendium of solution principles from the bioworld are needed.

Hopefully, dear reader, you will contribute.

7.9   REFERENCES

[1] J. Richardson, T. Irwin, and C. Sherwin, *Design and Sustainability: A Scoping Report for the Sustainable Design Forum*, Design Council, London, UK, 2005. www.designcouncil.org.uk/red 77

[2] World Commission on Environment and Development, *Our Common Future (The Brundtland Report)*. http://www.un-documents.net/wced-ocf.htm 77

[3] United Nations, *Sustainable Development Goals*. https://sustainabledevelopment.un.org/?menu=1300 77

[4] J. Rockström and P. Sukhdev, *How Food Connects All the SDGs*, Stockholm Resilience Centre, Stockholm, Sweden, 2019. https://www.stockholmresilience.org/research/research-news/2016-06-14-how-food-connects-all-the-sdgs.html 77

[5] S. Rahman, U. Sumotarto, and H. Pramudito, Influence the condition land subsidence and groundwater impact of Jakarta coastal area, *IOP Conference Series: Earth and Environmental Science*, 106:012006, 2018. DOI: 10.1088/1755-1315/106/1/012006. 79

[6] E. D. Ongley, Control of water pollution from agriculture, *Food and Agriculture Organization of the United Nations*, Rome, Italy, 1996. http://www.fao.org/3/w2598e/w2598e00.htm 79

[7] M. Z. Hauschild, J. Jeswiet, and L. Alting, Design for environment—Do we get the focus right? *CIRP Annals*, 53:1–4, 2004. DOI: 10.1016/s0007-8506(07)60631-3. 79

[8] Worldometers, *Live Counters of the World Population*. https://www.worldometers.info 79

[9] United Nations, *Sustainable Development Goal 1—End Poverty in All its Forms Everywhere*. https://sustainabledevelopment.un.org/sdg1 79

[10] World Bank, *Poverty—Overview*. https://www.worldbank.org/en/topic/poverty/overview 79

[11] T. A. Lenau and T. Hesselberg, Biomimetic self-organization and self-healing, *Engineered Biomimicry*, A. Lakhtakia and R. J. Martín-Palma, Eds., pages 333–358, Elsevier, Waltham, MA, 2013. DOI: 10.1016/b978-0-12-415995-2.00013-1 80

[12] L. Addadi, D. Joester, F. Nudelman, and S. Weiner, Mollusk shell formation: A source of new concepts for understanding biomineralization processes, *Chemistry: A. European Journal*, 12:980–987, 2006. DOI: 10.1002/chem.200500980. 80

[13] J. F. V. Vincent, O. A. Bogatyreva, N. R. Bogatyrev, A. Bowyer, and A.-K. Pahl, Biomimetics: Its practice and theory, *Journal of the Royal Society Interface*, 3:471–482, 2006. DOI: 10.1098/rsif.2006.0127. 81

[14] J. F. V. Vincent and U. G. K. Wegst, Design and mechanical properties of insect cuticle, *Arthropod Structure and Development*, 33:187–199, 2004. DOI: 10.1016/j.asd.2004.05.006. 81

[15] C. L. Craig, Evolution of arthropod silks, *Annual Review of Entomology*, 42:231–267, 1997. DOI: 10.1146/annurev.ento.42.1.231. 81

[16] D. B. Peakall, Conservation of web proteins in the spider, *Araneus diadematus*, *Journal of Experimental Zoology*, 176:257–264, 1997. DOI: 10.1002/jez.1401760302. 81, 89

[17] B. D. Opell, Economics of spider orb-webs: The benefits of producing adhesive capture thread and of recycling silk, *Functional Ecology*, 12:613–624, 1998. DOI: 10.1046/j.1365-2435.1998.00222.x. 81, 89

[18] R. F. Evert and S. E. Eichhorn, *Raven: Biology of Plants*, 8th ed., W. H. Freeman, New York, 2013. DOI: 10.1007/978-1-319-15626-8. 81, 82, 88

[19] D. W. Lee, *Nature's Palette: The Science of Plant Color*, University of Chicago Press, Chicago, IL, 2007. DOI: 10.7208/chicago/9780226471051.001.0001. 82

[20] G. Strout, S. D. Russell, D. P. Pulsifer, S. Erten, A. Lakhtakia, and D. W. Lee, Silica nanoparticles aid in structural leaf coloration in the Malaysian tropical rainforest understorey herb *Mapania caudata*, *Annals of Botany*, 112:1141–1148, 2013. DOI: 10.1093/aob/mct172. 82

[21] S. Kinoshita, *Structural Colors in the Realm of Nature*, World Scientific, Singapore, 2008. DOI: 10.1142/6496. 82

[22] N. Dushkina and A. Lakhtakia, Structural colors, *Engineered Biomimicry*, A. Lakhtakia and R. J. Martín-Palma, Eds., pages 267–303, Elsevier, Waltham, MA, 2013. DOI: 10.1016/c2011-0-06814-x. 82

[23] D. H. Evans, P. M. Piermarini, and K. P. Choe, The multifunctional fish gill: Dominant site of gas exchange, osmoregulation, acid-base regulation, and excretion of nitrogenous waste, *Physiological Reviews*, 85:97–177, 2005. DOI: 10.1152/physrev.00050.2003. 82

[24] A. Lakhtakia, From bioinspired multifunctionality to mimumes, *Bioinspired, Biomimetic and Nanobiomaterials*, 4:168–173, 2015. DOI: 10.1680/jbibn.14.00034. 82

[25] R. Sender, S. Fuchs, and R. Milo, Revised estimates for the number of human and bacteria cells in the body, *PLoS Biology*, 14:e1002533, 2016. DOI: 10.1371/journal.pbio.1002533. 82

[26] Y. E. Borre, R. D. Moloney, G. Clarke, T. G. Dinan, and J. F. Cryan, The impact of microbiota on brain and behavior: Mechanisms and therapeutic potential, *Microbial Endocrinology: The Microbiota-Gut-Brain Axis in Health and Disease*, M. Lyte and J. F. Cryan, Eds., pages 373–403, Springer, New York, 2014. DOI: 10.1007/978-1-4939-0897-4. 82

[27] E. van Nood, A. Vrieze, M. Nieuwdorp, S. Fuentes, E. G. Zoetendal, W. M. de Vos, C. E. Visser, E. J. Kuijper, J. F. W. M. Bartelsman, J. G. P. Tijssen, P. Speelman, M. G. W. Dijkgraaf, and J. J. Keller, Duodenal infusion of donor feces for recurrent *Clostridium difficile*, *New England Journal of Medicine*, 368:407–415, 2013. DOI: 10.1056/NEJMoa1205037. 82

[28] G. R. Jones and J. E. Parker, Pheromones, *Encyclopedia of Analytical Science*, 2nd ed., P. J. Worsfold, A. Townshend, and C. F. Poole, Eds., pages 140–149, Elsevier, Amsterdam, The Netherlands, 2005. 83

[29] P. M. Fallow, B. J. Pitcher, and R. D. Magrath, Alarming features: Birds use specific acoustic properties to identify heterospecific alarm calls, *Proceedings of the Royal Society of London B*, 280:20122539, 2013. DOI: 10.1098/rspb.2012.2539. 83

[30] I. Dahlin, L. P. Kiær, G. Bergkvist, M. Weih, and V. Ninkovic, Plasticity of barley in response to plant neighbors in cultivar mixtures, *Plant and Soil*, 447:537–551, 2020. DOI: 10.1007/s11104-019-04406-1. 83

[31] Symbiosis Center Denmark, *Dansk Symbiosecenter: The Potential is 10,000 Jobs*. https://symbiosecenter.dk 83

[32] Kalundborg Symbiose, *Partnership Between Public and Private Companies in Kalundborg*. http://symbiosis.dk 83

[33] S. Wittrup, Pig City: The piggery of the future will have a nursery on the roof, *Ingeniøren*, Danish Society of Engineers, Copenhagen, Denmark, 2010. https://ing.dk/artikel/pig-city-fremtidens-grisestald-far-gartneri-pa-forste-sal-105643 83

[34] J. M. Rayner, Avian flight dynamics, *Annual Review of Physiology*, 44:109–119, 1982. 84

[35] T. Alerstam, M. Rosén, J. Bäckman, P. G. P. Ericson, and O. Hellgren, Flight speeds among bird species: Allometric and phylogenetic effects, *PLoS Biology*, 5:e197, 2007. DOI: 10.1371/journal.pbio.0050197. 84

[36] T. J. Dawson, Kangaroos, *Scientific American*, 237(2):78–89, 1977. https://www.jstor.org/stable/24954004 84

[37] P. Ball, Shark skin and other solutions, *Nature*, 400:507–509, 1999. DOI: 10.1038/22883. 84, 88

[38] T. Sullivan and F. Regan, The characterization, replication and testing of dermal denticles of *Scyliorhinus canicula* for physical mechanisms of biofouling prevention, *Bioinspiration and Biomimetics*, 6:046001, 2011. DOI: 10.1088/1748-3182/6/4/046001. 84

[39] J. Davenport, R. N. Hughes, M. Shorten, and P. S. Larsen, Drag reduction by air release promotes fast ascent in jumping emperor penguins—a novel hypothesis, *Marine Ecology Progress Series*, 430:171–182, 2011. DOI: 10.3354/meps08868. 84, 88

[40] C. Telenko, *Developing green design guidelines: A formal method and case study*, Ph.D. Dissertation, University of Texas at Austin, Austin, TX, 2009. https://repositories.lib.utexas.edu/handle/2152/ETD-UT-2009-12-591 84

[41] J. M. O'Rourke and C. C. Seepersad, Toward a methodology for systematically generating energy- and materials-efficient concepts using biological analogies, *Journal of Mechanical Design*, 137:091101, 2015. DOI: 10.1115/1.4030877. 84

[42] ISO 18458:2015, *Biomimetics—Terminology, Concepts and Methodology*, International Standards Organization, Geneva, Switzerland, 2015. https://www.iso.org/standard/62500.html DOI: 10.3403/30274979. 85

[43] The Biomimicry Institute, *Biomimicry DesignLens: Life's Principles*, Missoula, MT. http://biomimicry.net/about/biomimicry/biomimicry-designlens/ 85

[44] The Biomimicry Institute, *AskNature: Innovation Inspired by Nature*, Missoula, MT. https://asknature.org/ 85

[45] T. A. Lenau, A. M. Orrú, and L. Linkola, *Biomimicry in the Nordic Countries*, Nordisk Ministerråd, Copenhagen, Denmark, 2018. https://doi.org/10.6027/NA2018-906 DOI: 10.6027/na2018-906. 85

[46] T. Mead and S. Jeanrenaud, The elephant in the room: Biomimetics and sustainability?, *Bioinspired, Biomimetic and Nanobiomaterials*, 6:113–121, 2017. DOI: 10.1680/jbibn.16.00012. 85

[47] Ellen Macarthur Foundation, *Towards the Circular Economy*, London, UK, 2013. https://www.ellenmacarthurfoundation.org/assets/downloads/publications/Ellen-MacArthur-Foundation-Towards-the-Circular-Economy-vol.1.pdf 86

[48] Ellen Macarthur Foundation, *A New Textiles Economy: Redesigning Fashion's Future*, London, UK, 2017. https://www.ellenmacarthurfoundation.org/publications/a-new-textiles-economy-redesigning-fashions-future 86

[49] S. Baughan, *What Happens to the Clothes that you Dispose of?*. https://www.loveyourclothes.org.uk/blogs/what-happens-clothes-you-dispose 86

[50] A. Bjørn and M. Z. Hauschild, Absolute versus relative environmental sustainability: What can the cradle-to-cradle and eco-efficiency concepts learn from each other?, *Journal of Industrial Ecology*, 17:321–332, 2013. DOI: 10.1111/j.1530-9290.2012.00520.x. 86, 87

[51] W. McDonough and M. Braungart, *Cradle to Cradle: Remaking the Way We Make Things*, North Point Press, New York, 2002. 86

[52] I. C. de Pauw, P. Kandachar, and E. Karana, Assessing sustainability in nature-inspired design, *International Journal of Sustainable Engineering*, 8:5–13, 2015. DOI: 10.1080/19397038.2014.977373. 87

[53] T. A. Lenau, D. C. A. Pigosso, T. C. McAloone, and A. Lakhtakia, Biologically inspired design for environment, *Proceedings of SPIE*, 11374:113740E, 2020. DOI: 10.1117/12.2558498. 88

[54] A. A. Berryman, The origins and evolution of predator-prey theory, *Ecology*, 73:1530–1535, 1992. DOI: 10.2307/1940005. 88

[55] G. Pahl, W. Beitz, J. Feldhusen, and K.-H. Grote, *Engineering Design: A Systematic Approach*, 3rd ed., Springer, London, UK, 2007. DOI: 10.1007/978-1-84628-319-2. 88

Authors' Biographies

TORBEN A. LENAU

Torben A. Lenau is an Associate Professor in design methodology, material selection, and biomimetics at the Department of Mechanical Engineering, Danmarks Tekniske Universitet. His research interests are creative methods in product design with focus on materials, manufacturing, and biomimetics (inspiration from nature). He has conducted a number of industrial case studies on how to integrate biomimetics in product development and has developed the biocards used to communicate design principles found in nature. Furthermore, he studies natural occurring photonic structures in order to develop new surface coatings based on structural colors.

AKHLESH LAKHTAKIA

Akhlesh Lakhtakia is Evan Pugh University Professor and the Charles Godfrey Binder (Endowed) Professor of Engineering Science and Mechanics at The Pennsylvania State University. He received his B.Tech. (1979) and D.Sc. (2006) degrees in Electronics Engineering from the Institute of Technology, Banaras Hindu University, and his M.S. (1981) and Ph.D. (1983) degrees in Electrical Engineering from the University of Utah. He was the Editor-in-Chief of the *Journal of Nanophotonics* from its inception in 2007–2013. He has been elected a Fellow of the American Association for the Advancement of Sciences, American Physical Society, Institute of Physics (UK), Optical Society of America, SPIE–The International Society for Optics and Photonics, Institute of Electrical and Electronics Engineers, Royal Society of Chemistry, and Royal Society of Arts. His current research interests include: electromagnetic fields in complex mediums, sculptured thin films, mimumes, surface multiplasmonics and electromagnetic surface waves, forensic science, and engineered biomimicry.

Printed in the United States
by Baker & Taylor Publisher Services